Certified Wireless Specialist (CWS)

Study Guide

(CWS-100)

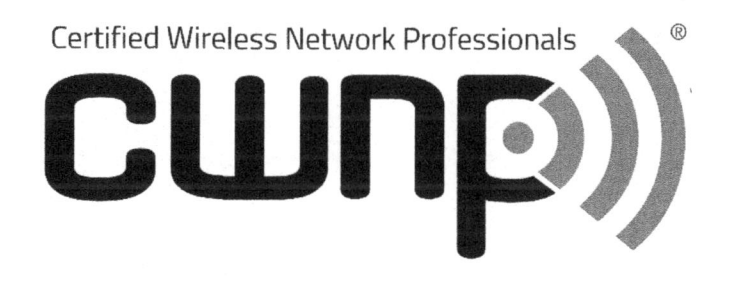

Certified Wireless Network Professionals

cwnp

Copyright © 2017 by CertiTrek Publishing. All rights reserved. Printed in the United States of America. Except as permitted under the United States Copyright Act of 1976, no part of this publication may be reproduced or distributed in any form or by any means, or stored in a database or retrieval system, without the prior written permission of the publisher.

All trademarks or copyrights mentioned herein are the possession of their respective owners, and CertiTrek Publishing makes no claim of ownership by the mention of products that contain these marks.

Errata, when available, for this study guide, can be found at www.cwnp.com/errata/

First printing December 2017, version 1.0

ISBN: 9781717720993

Author
Tom Carpenter

Technical Editor
Manon Lessard

Editor
Sean Stallings

Table of Contents

Introduction	iv
Extended Table of Contents	xii
Chapter 1 – Radio Frequency (RF)	1
Chapter 2 – RF Hardware	35
Chapter 3 – The 802.11 Standard	59
Chapter 4 – Channel Plans and Performance Factors	85
Chapter 5 – WLAN Security, BYOD and Guest Access	115
Chapter 6 – Enhanced 802.11 Functions	145
Chapter 7 – Wireless Access Points	166
Chapter 8 – Wireless clients	188
Chapter 9 – WLAN Requirements	208
Glossary: A CWNP Universal Glossary	236

Introduction

The Certified Wireless Specialist (CWS) is an individual who understands the fundamentals of Wi-Fi networks and can recommend or select the best solutions for a given installation specification. The CWS-100 exam tests the candidate's knowledge to verify his or her ability to perform the duties of a CWS.

The CWS is an individual who can explain essential features and capabilities for wireless Local Area Network (WLAN) solutions including Access Points (APs), controllers, WLAN management solutions, and 802.11 networks. The individual can assist in selecting the best equipment for a deployment or communicate well with those who are responsible for such decisions. The CWS is not responsible for configuration and management of the WLAN, but must have the ability to gather information used to determine requirements and match technologies to those requirements for a deployment.

The CWS-100 exam consists of 60 multiple choice, single correct answer questions and is delivered online by CWNP. The exam is purchased at CWNP.com. The candidate is given 90 minutes to take the exam and must achieve a score of 70% or higher to earn the CWS certification.

Book Features

The CWS Study guide includes the following features:

- End of chapter Points to Remember. These lists of important facts help you retain the information learned in the chapter.

- End of chapter review quizzes. These quizzes help you test the knowledge you have acquired from the chapter. Each chapter contains 10 quiz items.

- Notes with special indicators. The notes throughout the book fall into one of three categories as outlined in Table i.1.

- CWNP official glossary. A glossary of terms provided at the end of the book that helps you as a reference while reading.

- Full-color printing. All screenshots and diagrams are created in color to assist in the learning process and provide detailed and relevant information in the images.

- Complete coverage of the CWS-100 objectives. Every objective is covered in the book, and each chapter lists the objectives covered within.

Icon	Description
	Note: A general note related to the current topic.
	Defined Note: A note providing a concise definition of a term or concept.
	Exam Note: A note providing tips for exam preparation.

Table i.1: Book Note Icons

About the Author

Tom Carpenter is the CTO of CWNP and has more than 20 years of experience in the Information Technology industry. He has written 18 previous books and developed more than 50 eLearning programs in the past fifteen years. He is a CWNE and holds several other industry certifications as well. As the CTO of CWNP, Tom is responsible for setting the direction for certifications and managing product development projects through their lifecycles. He can be reached at tom@cwnp.com and is heard on a monthly webinar series presented by CWNP and archived on YouTube on the CWNPTV channel. Tom lives in Ohio with his wife Julie and loves books and all things tech.

About the Technical Editor

Manon "Mae" Lessard is a Canadian network administrator for a large university. With 15 years experience in networking, 13 of which have been mostly spent exploring her passion for all things wireless, she is an active member of the Wireless community. She is involved in various organizations such as the Wireless LAN Association and the Slack WLAN-Pro community, holds several industry certifications and is currently working towards earning her CWNE. You can reach her on Twitter (@Mae149) or through her blog at missmaeswifi.com.

CWS-100 Objectives

The CWS-100 exam tests your knowledge against four knowledge domains as documented in Table i.2. The CWS candidate should understand these domains before taking the exam. The CWS-100 objectives follow.

Knowledge Domain	Percentage
Understand Basic RF Hardware and Functions	15%
Identify 802.11 Features and Functions	30%
Identify Wireless LAN Hardware and Software	30%
Understand Organizational Goals	25%

Table i.2: CWS-100 Exam Knowledge Domains with Percentage of Questions in Each Domain

Understand Basic RF Hardware and Functions (15%)

1.1 Identify RF characteristics
- 1.1.1 RF waves
- 1.1.2 Amplitude
- 1.1.3 Frequency
- 1.1.4 Wavelength

1.2 Explain basic RF behaviors
- 1.2.1 Reflection
- 1.2.2 Absorption
- 1.2.3 Signal strength

1.3 Understand antenna types
- 1.3.1 Omnidirectional
- 1.3.2 Semi-directional
- 1.3.3 Highly directional
- 1.3.4 Internal vs. external

Identify 802.11 Features and Functions (30%)

2.1 Know the frequency bands used

      2.1.1   2.4 GHz – 802.11b/g/n
      2.1.2   5 GHz – 802.11a/n/ac
      2.1.3   Sub-1 GHz – 802.11ah
      2.1.4   60 GHz – 802.11ad
2.2 Identify Physical Layer (PHY) characteristics
      2.2.1   Data rates
      2.2.2   Bands used
      2.2.3   Supported technologies (laptops, tablets, video devices, Internet of Things (IoT))
2.3 Select appropriate channels
      2.3.1   Channel selection best practices
      2.3.2   Common channel selection mistakes
2.4 Identify factors impacting wireless LAN (WLAN) performance
      2.4.1   Coverage requirements
      2.4.2   Capacity requirements
      2.4.3   Required features
      2.4.4   Poor configuration and implementation
2.5 Explain the basic differences between WPA and WPA2 security
      2.5.1   Authentication and key management
      2.5.2   Encryption
      2.5.3   Personal vs. Enterprise
2.6 Describe features of enhanced 802.11 functions
      2.6.1   Mesh
      2.6.2   Quality of Services (QoS)
      2.6.3   SISO vs. MIMO
      2.6.4   Dynamic Rate Switching (DRS)
      2.6.5   Backwards compatibility

Identify Wireless LAN Hardware and Software (30%)

3.1 Identify AP features and capabilities
      3.1.1   PHY support
      3.1.2   Single-band vs. dual-band
      3.1.3   Output power control

- 3.1.4 Operational modes
- 3.1.5 Multiple-SSID support
- 3.1.6 Guest access
- 3.1.7 Security features
- 3.1.8 Management interfaces
- 3.1.9 Internal and external antennas
- 3.1.10 PoE support

3.2 Describe AP management systems
- 3.2.1 Autonomous
- 3.2.2 Controller
- 3.2.3 Cloud
- 3.2.4 Management systems

3.3 Determine capabilities of client devices
- 3.3.1 PHY support
- 3.3.2 Single-band vs. multi-band
- 3.3.3 Support for MIMO
- 3.3.4 Supported channels in 5 GHz
- 3.3.5 Supported security options

3.4 Identify when Power over Ethernet (PoE) should be used

3.5 Explain the requirements of fast and secure roaming for non-technical professionals
- 3.5.1 Latency requirements for streaming communications
- 3.5.2 Pre-authentication
- 3.5.3 Key caching methods

3.6 Understand the basic requirements for voice over WLAN (VoWLAN)
- 3.6.1 Latency
- 3.6.2 Jitter
- 3.6.3 Signal strength

3.7 Determine the best solution for BYOD and guest access
- 3.7.1 User provisioning
- 3.7.2 Captive portals
- 3.7.3 Device and software control solutions

Understand Organizational Goals (25%)

4.1 Understand issues in common vertical markets
- 4.1.1 Standard Enterprise Offices
- 4.1.2 Healthcare
- 4.1.3 Hospitality
- 4.1.4 Conference Centers
- 4.1.5 Education
- 4.1.6 Government
- 4.1.7 Retail
- 4.1.8 Industrial
- 4.1.9 Emergency Response
- 4.1.10 Temporary Deployments
- 4.1.11 Small Office/Home Office (SOHO)
- 4.1.12 Public Wi-Fi

4.2 Gather information about existing networks
- 4.2.1 Network diagrams
- 4.2.2 Wi-Fi implementations
- 4.2.3 Neighbor networks
- 4.2.4 Available network services
- 4.2.5 PoE availability

4.3 Discover coverage and capacity needs
- 4.3.1 Define coverage areas
- 4.3.2 Define capacity zones

4.4 Discover client devices and applications in use
- 4.4.1 Laptops, tablets, mobile phones, desktops, and specialty devices
- 4.4.2 Real-time applications
- 4.4.3 Standard applications (e-mail, web browsing, database access, etc.)
- 4.4.4 Data-intensive applications (file downloads/uploads, cloud storage, cloud backup, etc.)

4.5 Determine the need for outdoor coverage networks and bridge links
- 4.5.1 Bridge link distance and required throughput

    4.5.2   Outdoor areas requiring coverage
    4.5.3   Use cases for outdoor access
4.6 Define security constraints
    4.6.1   Regulatory
    4.6.2   Industry standards and guidelines
    4.6.3   Organizational policies
4.7 Discover use cases and access types
    4.7.1   Authorized users
    4.7.2   Onboarded guest access
    4.7.3   Public Wi-Fi
4.8 Match organizational goals to WLAN features and functions

Each chapter of the Certified Wireless Specialist (CWS) Official Study Guide lists the objectives covered in that chapter on the chapter title page.

Extended Table of Contents

Table of Contents --- iii

Introduction --- iv

    Book Features --- v

    About the Author --- vi

    About the Technical Editor --- vi

    CWS-100 Objectives --- vii

    Understand Basic RF Hardware and Functions (15%) --- vii

    Identify 802.11 Features and Functions (30%) --- vii

    Identify Wireless LAN Hardware and Software (30%) --- viii

    Understand Organizational Goals (25%) --- x

Extended Table of Contents --- xii

Chapter 1 – Radio Frequency (RF) --- 1

    Objectives Covered --- 1

    Understanding Radio Frequency --- 2

    RF Characteristics --- 5

    RF Behaviors --- 11

    How Wi-Fi Uses RF --- 26

    Chapter Summary --- 30

    Points to Remember --- 30

    Review Questions --- 31

    Review Answers --- 33

Chapter 2 – RF Hardware --- 35

    Objectives Covered --- 35

    Understanding Antennas --- 36

Antenna Types ---- 44
Internal vs. External Antennas ---- 53
Additional RF Hardware ---- 54
Chapter Summary ---- 55
Points to Remember ---- 55
Review Questions ---- 56
Review Answers ---- 58

Chapter 3 – The 802.11 Standard ---- 59
Objectives Covered ---- 59
IEEE Standards ---- 60
Frequency Bands Used by 802.11 Devices ---- 65
Physical Layers in 802.11 ---- 66
Key 802.11 Standard Non-PHY Amendments ---- 76
802.11 Networking Terminology ---- 78
Chapter Summary ---- 81
Points to Remember ---- 81
Review Questions ---- 82
Review Answers ---- 84

Chapter 4 – Channel Plans and Performance Factors ---- 85
Objectives Covered ---- 85
WLAN Performance Factors ---- 86
Channel Selection ---- 97
Voice over WLAN (VoWLAN) Requirements ---- 103
Chapter Summary ---- 111
Points to Remember ---- 111

Review Questions	112
Review Answers	114 |//
Chapter 5 – WLAN Security, BYOD, and Guest Access	**115**
Objectives Covered	115
WLAN Security Basics	116
BYOD	132
Guest Access	137
Chapter Summary	140
Points to Remember	140
Review Questions	141
Review Answers	143
Chapter 6 – Enhanced 802.11 Functions	**145**
Objectives Covered	145
Enhanced Features of 802.11	146
Fast Secure Roaming (FSR)	157
Chapter Summary	161
Points to Remember	161
Review Questions	162
Review Answers	164
Chapter 7 – Wireless Access Points	**166**
Objectives Covered	166
AP Features and Capabilities	167
AP Management Solutions	179
Chapter Summary	183
Points to Remember	183

Review Questions --- 184
Review Answers --- 186
Chapter 8 – Wireless clients --- 188
Objectives Covered --- 188
Client Device Types --- 189
Single-Band vs. Multi-Band --- 195
Physical Layers (PHY) Supported --- 197
Supported Channels --- 198
Supported Security Options --- 200
Applications in Use --- 201
Chapter Summary --- 203
Points to Remember --- 203
Review Questions --- 204
Review Answers --- 206
Chapter 9 – WLAN Requirements --- 208
Objectives Covered --- 208
Common Vertical Markets --- 209
Gathering Essential Information --- 216
Chapter Summary --- 230
Points to Remember --- 230
Review Questions --- 231
Review Answers --- 234
Glossary: A CWNP Universal Glossary --- 236
Index --- 266

Chapter 1 – Radio Frequency (RF)

Objectives Covered

1.1 Identify RF characteristics
- 1.1.1 RF waves
- 1.1.2 Amplitude
- 1.1.3 Frequency
- 1.1.4 Wavelength

1.2 Explain basic RF behaviors
- 1.2.1 Reflection
- 1.2.2 Absorption
- 1.2.3 Signal strength

Radio frequency is the foundation of Wi-Fi networks. A basic understanding of radio frequencies is required to define, select and explain Wi-Fi hardware and software to others. As a sales professional, it is essential to understand these concepts. As a manager, project manager or consultant, it is necessary to select the best solutions for your organization. This knowledge also helps students to learn more advanced concepts covered in the Certified Wireless Network Administrator (CWNA) study materials and exam as well as the Certified Wireless Technologist exam, which is focused more on the installation and configuration of necessary wireless hardware and software. Therefore, this chapter provides the foundation for understanding all of the remaining concepts in this Certified Wireless Specialist (CWS) study guide.

Understanding Radio Frequency

Radio Frequency (RF) is a term used to reference a portion of the electromagnetic spectrum that is used for 802.11 (and other) network communications. Wi-Fi networks use RF waves in the microwave frequency range. Microwave frequencies range from 300 megahertz (MHz) to 300 gigahertz (GHz), which represents wavelengths (covered later in this chapter) from 1 meter to 1 millimeter (or from 100 centimeters to 0.1 centimeters). Notice that the higher frequencies result in shorter wavelengths, which is knowledge useful later in this chapter. Frequencies used in 802.11 networks range from 700 MHz to 60 GHz. The vast majority of Wi-Fi networks use the 2.4 GHz and 5 GHz frequency bands. Figure 1.1 shows the electromagnetic spectrum, including the visible light region, since light is an electromagnetic wave as well.

> *RF waves* are used to carry information in wireless connections. They are electromagnetic waves existing in the microwave range of the electromagnetic spectrum shown in Figure 1.1. The RF waves are manipulated (modulated) in different ways to represent digital data bits. They have characteristics that define them and behaviors that impact how they travel through free space, office environments, and other environments.

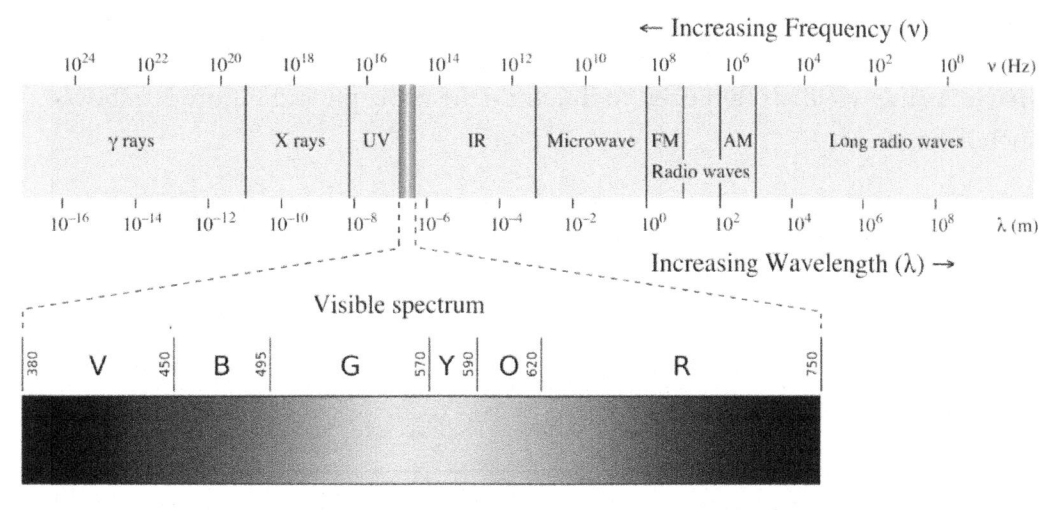

Figure 1.1: Electromagnetic Spectrum

RF is the carrier used to *modulate* data. Modulation is the process used to manipulate RF waves so that they represent digital data bits. The RF wave amplitude (strength) and phase (relation to other waves), which is defined in detail later in this chapter, can be changed to indicate binary data. 802.11 networks use several different modulation techniques, including Binary Phase Shift Keying (BPSK), Quadrature Phase Shift Keying (QPSK), 16-Quadrature Amplitude Modulation (QAM), 64-QAM and 256-QAM. The modulation technique used, in addition to coding rates and a few other factors, determines the data rate for a given channel width (range of frequencies used to define the channel). When an RF wave is modulated, it is known as a *carrier wave*, and it represents a *signal*. These concepts are explored in more detail throughout this chapter.

> A *channel* is defined by its width and is referenced as a number. The width is the range of frequencies (the rate at which an RF wave repeats) used in the channel. The standard (Institute of Electrical and Electronics Engineers (IEEE) 802.11) defines the channel numbers.

A wave, whether within the RF range or otherwise, can be defined as a motion or oscillation that travels through a medium. The medium can be a mass or free space. A sine wave is often used to illustrate the concept, and Figure 1.2 shows such a wave.

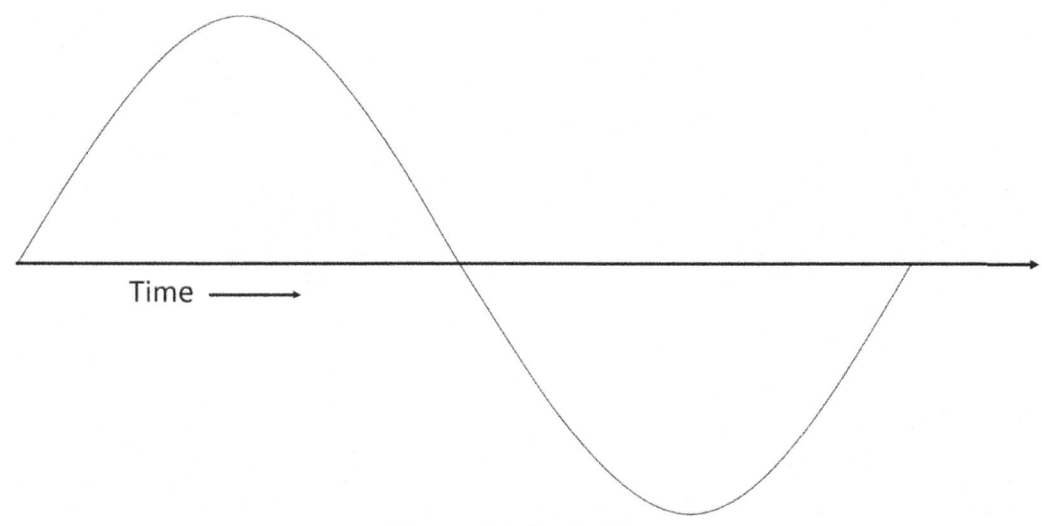

Figure 1.2: Basic Wave

RF waves are generated both intentionally and incidentally. Intentional waves are generated by radiators that provide RF-based communications. Incidental waves are those generated by electric motors and other devices that are not intended to act as signals. When selecting and installing Wi-Fi equipment, it is important to remember that other RF generators (both intentional and incidental) may cause interference with the Wi-Fi network.

> Wi-Fi networks are known as wireless local area networks (WLANs) in the engineering community. This book uses the terms Wi-Fi, wireless network, and WLAN interchangeably.

When RF waves are used to transmit signals, they are considered the medium used for transmission. Sometimes the phrase *RF medium* is used to refer to these waves that are modulated to carry data.

RF Characteristics

While RF waves have many properties or characteristics that are important to understand, a CWS only needs to be familiar with simple characteristics, including frequency, wavelength, phase, and amplitude.

Frequency

The *frequency* of a wave is the rate at which it repeats or oscillates. A higher frequency wave cycles more often, and a lower frequency wave cycles less often. Also, longer wavelengths correspond to lower frequency waves, and shorter wavelengths correspond to higher frequency waves. Figure 1.3 shows a lower and higher frequency wave representation. The entire cycle of the wave (the upward crest and the downward crest) represents the entire waveform. If this waveform repeats 100 times per second, the frequency is 100 times per second.

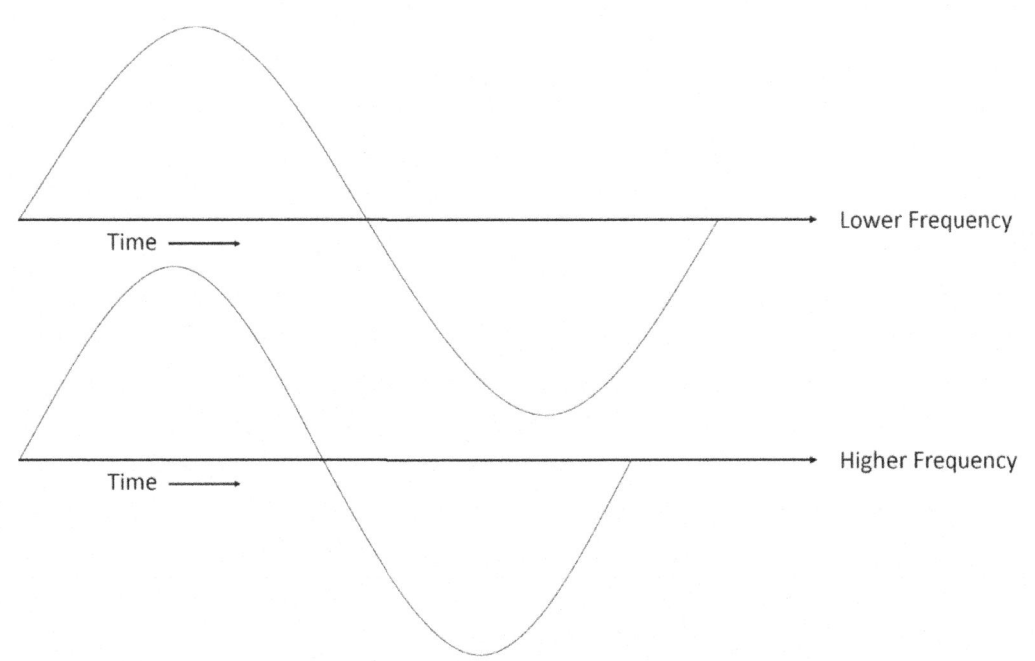

Figure 1.3: Varying Frequencies Illustrated

The frequency of a wave is designated in *Hertz*, named after Heinrich Hertz, who was a physicist living from 1857 to 1894. He proved the existence of electromagnetic waves, which were first mathematically theorized by James Clerk Maxwell. Frequency is measured in Hertz (Hz), kilohertz (kHz), megahertz (MHz) and gigahertz (GHz) within the ranges used by RF communication devices. One Hertz is equal to one wave cycle per second. Therefore, one kHz is 1000 Hertz, one MHz is 1,000,000 (one million) Hertz, and one GHz is 1,000,000,000 (one billion) Hertz.

IEEE 802.11 devices operate in four primary areas—sub-1 GHz, 2.4 GHz, 5 GHz and 60 GHz—and the vast majority of Wi-Fi devices operate within the frequency bands of 2.4 and 5 GHz. For example, laptops, tablets, mobile phones and most wireless desktops use 2.4 and 5 GHz frequencies. These ranges of use are called *bands* because they have a lower and upper range of utilization. For Wi-Fi, the 2.4 GHz band is from 2.400 GHz to 2.500 GHz, though 802.11 devices operate with primary frequencies used only from 2.401 to 2.495. Also, some of that space is not available worldwide due to constraints imposed by regulatory agencies (like the Federal Communications Commission (FCC)).

Modulation may use the frequency of the waves. This kind of modulation is called Frequency Shift Keying (FSK). FSK is not a modulation used within current 802.11 production devices, but you are likely familiar with it from Frequency Modulation (FM) radio.

Wavelength
With a basic understanding of frequency and its representation shown in Figure 1.3, the wavelength is easier to understand. The wavelength, though you cannot see RF waves with the human eye, is measured in actual distance in the same way as we measure physical objects, such as boards, clothing, and other materials. As stated previously, the wavelength is shorter for higher frequency waves and longer for lower frequency waves. The *wavelength* is defined as the distance required to complete a full wave cycle. Figure 1.4 illustrates this definition more clearly. Notice that the length of the wave is an entire cycle of the wave. Two cycles, and therefore two wavelengths, are represented in Figure 1.4.

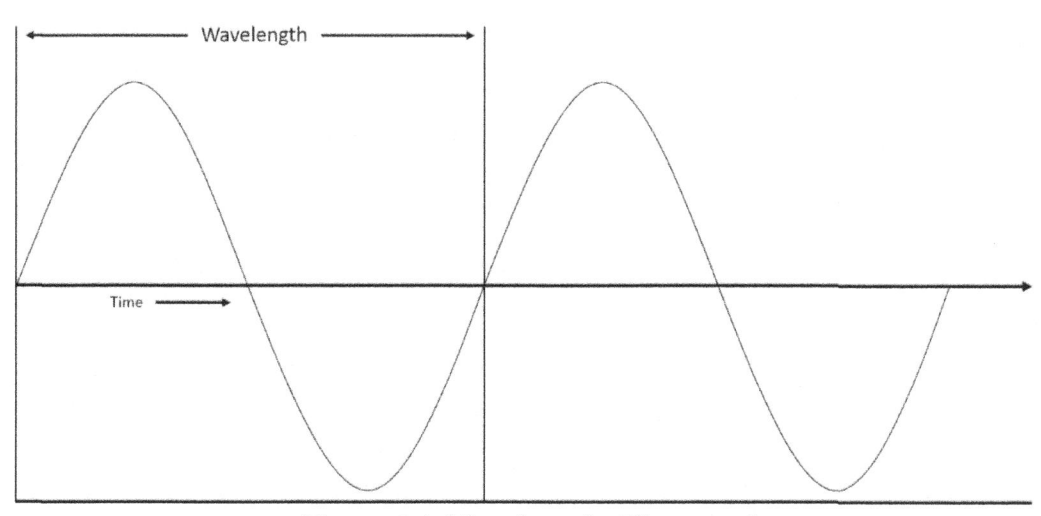

Figure 1.4: Wavelengths Illustrated

The wavelengths of the frequencies used for transmitting and receiving RF signals are critical. These wavelengths impact engineering and design decisions when creating antenna elements. Antennas must be designed with an understanding of the wavelength to optimize communications and maintain consistency.

As an example, 2.4 GHz RF waves are approximately 12.5 centimeters or 4.92 inches, and 5 GHz RF waves are approximately 6 centimeters or 2.36 inches. This wavelength variance is part of the reason that 5 GHz signals are not as easily received at greater distances when compared to 2.4 GHz signals. Designing antennas to receive weak RF signals becomes more and more challenging as the frequency increases and the wavelength decreases.

Figure 1.5 Further illustrates the varying wavelengths throughout the electromagnetic spectrum. Note that the image shows lower frequencies and longer wavelengths on the left and the higher frequencies and shorter wavelengths on the right.

Light waves are part of the same electromagnetic spectrum as RF waves. This fact becomes more important as we explore RF behaviors later in this chapter. Light waves fall within the frequency range of 430 to 750 terahertz (THz), which is well above the frequencies used for Wi-Fi networks.

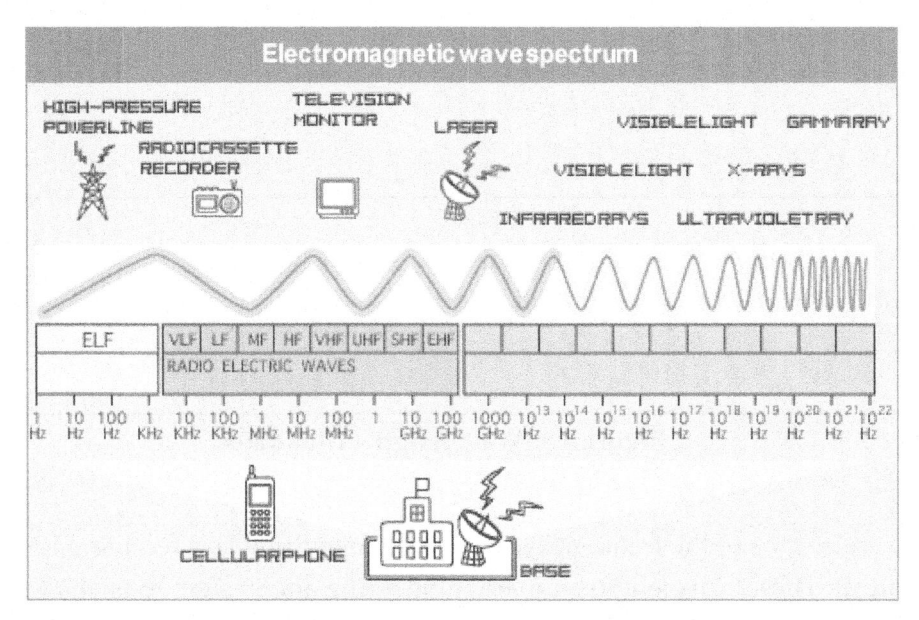

Figure 1.5: Wavelengths Represented across the Electromagnetic Spectrum

> Wavelengths even vary within bands. Different channels within the 2.4 GHz band have different wavelengths. The lengths stated in this section are estimates for the bands and are not exact. It is not essential to know exact wavelengths; instead, it is only essential to understand that they are longer at lower frequencies and shorter at higher frequencies.

Remember that Wi-Fi networks use frequencies that range from roughly 700 MHz to 60 GHz (with some possible connections using frequencies even lower than 700 MHz, such as TV whitespace 802.11 links). The GHz frequencies operate at a rate of the stated rate times one billion per second—for example, 5 GHz, is five times one billion or five billion times per second. THz frequencies operate at a rate of one trillion times per second times the stated rate. Therefore, light waves operate between 430 and 750 trillion times per second and have extremely short wavelengths.

Because of their short wavelengths, light waves may be impacted more than microwaves by specific materials. The most obvious example is that light waves do not penetrate solid materials (like walls and doors), but microwaves do. For this reason, while light waves are a tremendous thinking tool to learn about RF behaviors, they must not be considered synonymous with the RF waves that are used in Wi-Fi networks.

The wavelength is also minimally related to data rates. Higher frequency signals (with shorter wavelengths) can support higher data rates. For example, 2.4 GHz and 5 GHz networks support higher data rates than 900 MHz networks.

Phase

The *phase* of an RF wave is not a characteristic of an individual wave, but instead it is a comparison with another wave. A wave can either be 'in phase' with another wave or it can be, to some degree, out of phase with another wave. Phase can be used for modulation; for instance, the phase of a wave can be intentionally changed, as shown in Figure 1.6. Phase-based modulation is called Phase Shift Keying (PSK). Additionally, multiple copies of a signal can arrive at a receiver due to reflections as waves travel through environments (as you learn later in this chapter). If these copies arrive significantly out of phase, they can cause corruption or cancellation of the signal. This means that the receiver may not be able to process the signal correctly or at all.

Amplitude

The final RF wave characteristic that is important for the CWS candidate to understand is the amplitude. The *amplitude* of a wave is the power or strength of the wave. When an RF signal is received with higher amplitude, it is said to be a stronger signal; when it is received with a lower amplitude, it is said to be a weaker signal. The strength of the received signal is a factor of its amplitude at transmission and any effects that the wave experiences as it travels through the environment. An RF signal weakens as it travels through free space, and it weakens significantly more as it travels through materials, such as walls, doors, and floors.

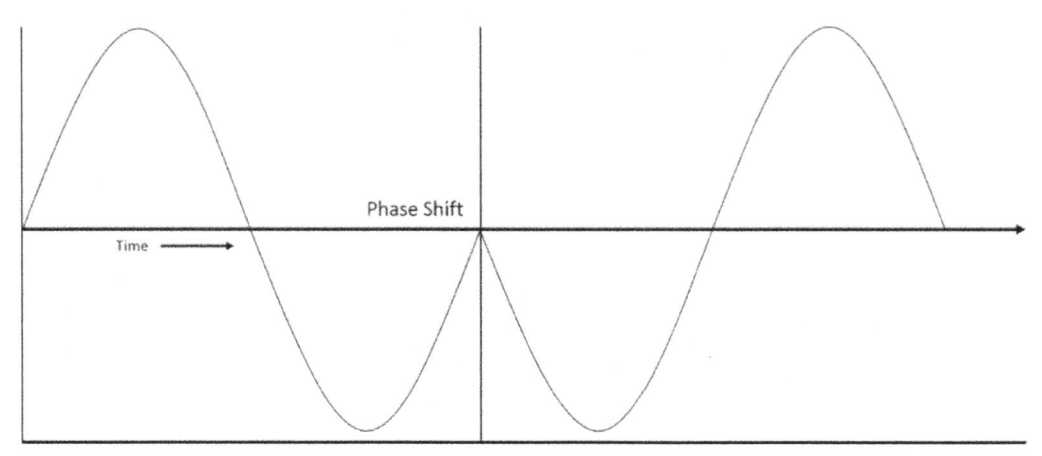

Figure 1.6: Phase Shifting Used for Modulation

Figure 1.7 shows an RF wave and an RF wave that has a reduced amplitude. A tool like a spectrum analyzer can be used to measure the strength of the RF energy at a given frequency and a given location. Like the frequency or phase of a wave, the amplitude of a wave can be used to modulate information onto a wave, making it a signal. Amplitude-based modulation is called Amplitude Shift Keying (ASK).

Both PSK and ASK are used in 802.11 production devices. By using more than one modulation technique at a time, you can achieve higher data rates. The lowest data rates in 802.11 devices use only PSK modulation. The higher data rates use a combination of PSK and ASK.

This section has introduced the basic concepts required of the CWS candidate and provides a foundation for the remaining sections of this chapter and the remainder of the book. With a basic understanding of RF waves, frequencies, wavelengths, phases, and amplitudes, you are prepared to understand the basics of RF behaviors and how Wi-Fi uses RF for communications. The remaining sections of this chapter explain these two crucial issues.

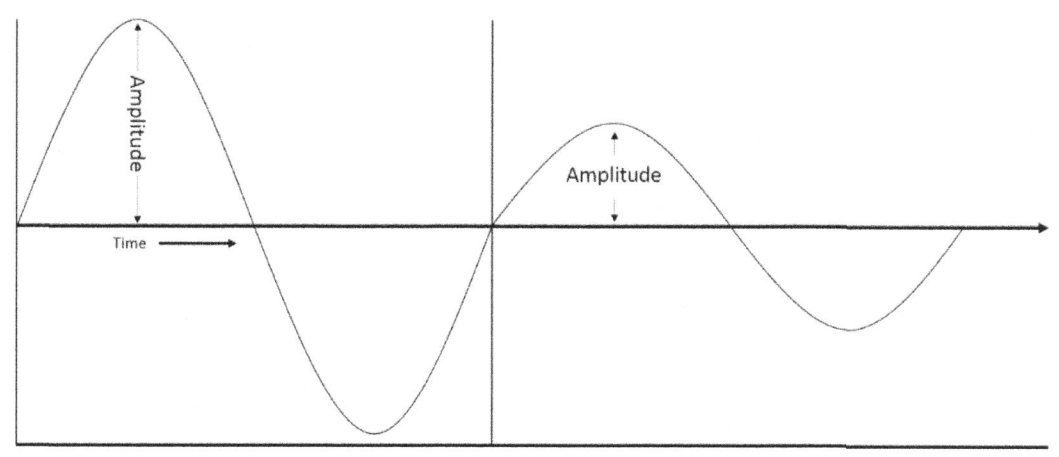

Figure 1.7: RF Wave Amplitude Represented

RF Behaviors

While we cannot see RF waves propagating (moving) through environments, we know certain things about their behaviors based on experiments and actual observation of light waves. Remember, light waves are part of the electromagnetic spectrum. We can see them and therefore can learn a great deal about their behavior. Through experimentation with RF waves, we can also determine that they behave in a similar manner.

> As stated in the preceding section, light waves form a good analogy to understand RF waves, but they are not identical to RF waves. While they can be used as a learning tool for understanding RF behavior, the student should never assume they are equal. Several differences exist, but common behaviors are also seen and these commonalities can assist the student when learning about RF waves.

In this section, you explore three key RF behaviors and how they impact signal strength at the receiver. These behaviors include reflection, attenuation, and absorption.

Reflection

Reflection occurs when an RF wave encounters an object that is smooth and larger than the length of the wave. The incoming wave is called the incident wave, and the reflected wave is called the reflected wave. Figure 1.8 illustrates the behavior of reflection.

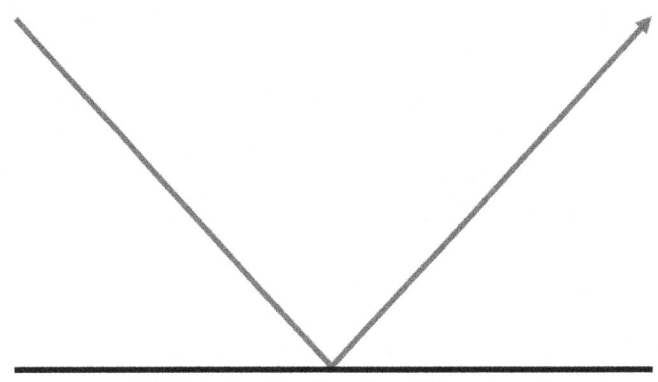

Figure 1.8: Reflection Illustrated

Reflection can be understood with the analogy of light. Consider, first, that what you see when you look at an object is light waves reflecting off of the object. Now, with that knowledge, consider a mirror. You may have noticed that you can look into a mirror at different angles and see different things reflected in the mirror. In fact, if you stand on the right side of the mirror, you can see objects that are actually on the left side of it and vice versa. This phenomenon occurs because you are positioning your eyes in the path of different reflected light waves as you move.

You can also see the reflection of light waves by shining a flashlight into a mirror and then changing the angle of the incident light beam. As you change the incident angle, you see the light reflected in a different direction from the mirror. While you can physically see the light waves in these examples, and the light waves reflect more easily off of much smaller objects (because light waves are so small themselves), RF waves reflect similarly.

The reflection of RF waves is usually a positive behavior today in indoor WLANs because it helps to provide coverage throughout a facility. Without reflections, many dead areas (areas without sufficient coverage) would exist. In a very highly reflective environment, early Wi-Fi communications experienced problems such as multipath effects (waves arriving at the receiver after traveling multiple reflected paths, resulting in weaker signals or nulled signals). When older Wi-Fi devices (e.g., 802.11a, 802.11g, 802.11b) experienced multipath effects, these effects caused degraded communications. Antenna diversity was used to increase the probability of countering its effects. Antenna diversity uses more than one antenna at the receiver and selects the antenna with the best incoming signal based on the first part of the signal, which is called the preamble and is transmitted before the actual data.

> When RF waves reflect, causing multiple signals to arrive at the receiver, this is known as *multipath*. The RF waves travel multiple paths (different paths) but arrive at the same receiver. Multipath can be a problem for some systems but is used as an advantage in modern 802.11n and 802.11ac systems.

Modern Wi-Fi devices (e.g., 802.11n, 802.11ac) use a technology called multiple input/multiple output (MIMO) that takes advantage of multipath effects to transmit multiple streams of data. The result is higher data rates, but, interestingly, it depends on reflections in the environment to achieve effective MIMO links. Therefore, what was once seen only as a negative behavior (i.e., multipath effects from reflections) is no longer considered a benefit in indoor WLANs.

Attenuation

An RF wave experiences *attenuation* after it leaves the transmitter. The wave may travel along an RF cable, through space with obstacles or through free space. In all cases, the wave strength or amplitude attenuates as it travels. This statement means that the amplitude of the wave is decreased.

Figure 1.9 illustrates the attenuation that may occur as an RF wave travels through a dense material, such as a wall, door or floor. The material absorbs some of the RF energy (absorption is covered in more detail later), and this results in higher attenuation than the wave would experience in free space.

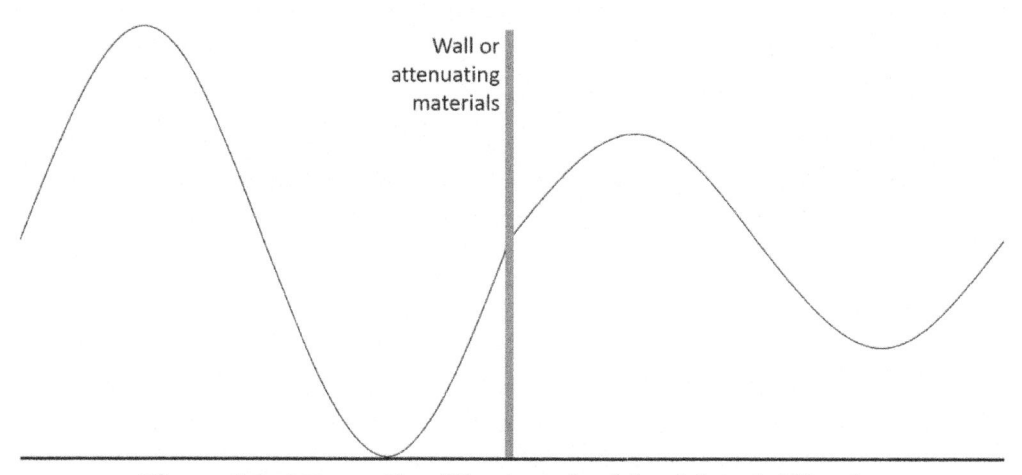

Figure 1.9: Attenuation Illustrated with a Material Barrier

Figure 1.10 illustrates attenuation in free space and is generated using iBwave Wi-Fi, which is a WLAN design tool. Given that the entire width of the predicted area (i.e., the area where RF signal strength is simulated) is 1000 meters, the RF waves attenuate quickly, even in free space. Within 500 meters, the signal strength has attenuated to the point that even the lowest of data rates would no longer function well for WLAN connectivity, given that each data rate requires a particular signal strength range. In reality, even at 300 meters, most WLAN designers would not accept the quality of the signal for typical use scenarios. At 300 meters, in this image, the average signal strength is around -85 to -88 dBm (dBm power levels are discussed later, but, for now, they are absolute power levels). This power level is usually only useful for 1 or 2 megabits per second (Mbps) data rates in Wi-Fi links at 2.4 GHz. Many engineers disable these legacy data rates (1 and 2 Mbps) that were prominently used in the 1990s. Figure 1.10 shows the coverage provided by channel 1 at 2.4 GHz.

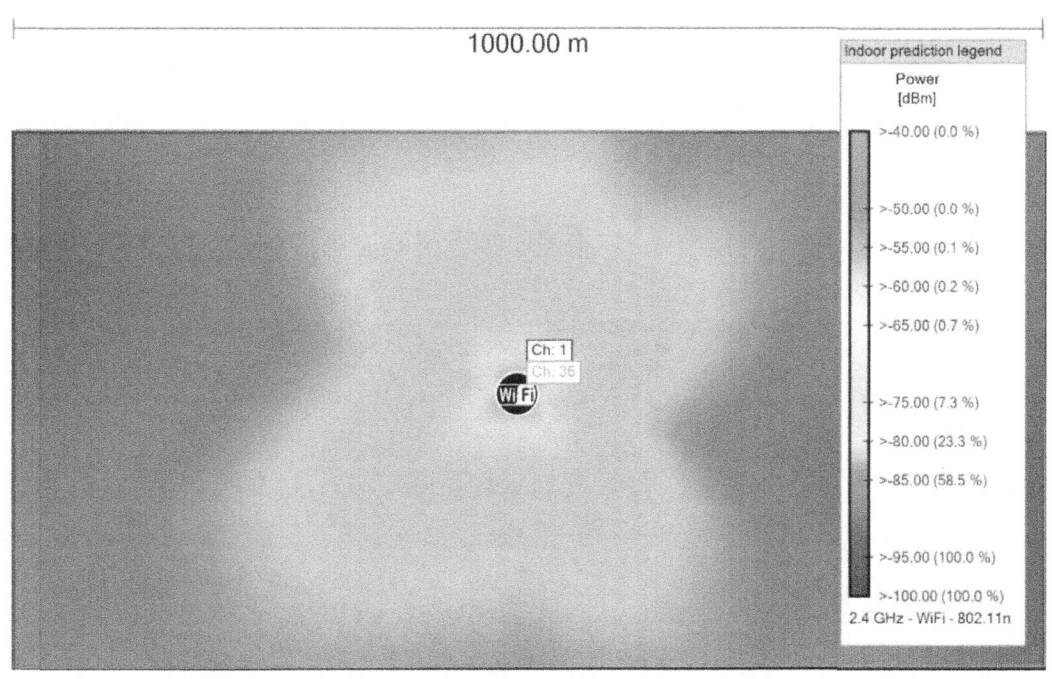

Figure 1.10: Attenuation at 2.4 GHz Illustrated in Free Space – Image Generated by iBwave Wi-Fi

Figure 1.11 is the same propagation model that is shown in Figure 1.10, but it shows the results for channel 36 at 5 GHz. Notice that the signal attenuates more quickly due to its shorter wavelength. At the same point of approximately 300 meters, the signal strength is around -88 dBm to -91 dBm, which is about half the power of the 2.4 GHz signal at the same location.

The fact that 5 GHz signals attenuate roughly twice as fast as 2.4 GHz signals in free space has led to two common recommendations in order to achieve coverage and capacity in a deployment. One recommendation suggests turning off the 2.4 GHz radio in many of your APs. The other suggests increasing the output power on the 5 GHz radios by at least 3 dB. The 2.4 GHz radios are turned off because fewer radios cover more space. Alternatively, you can double the power in the 5 GHz radios to gain a more balanced configuration. However, every space is different, and in some spaces, regardless of our desire for more 5 GHz clients, many more 2.4 GHz clients must be supported. Again, more information about dBm and dB is covered later in this chapter. More information about channel selection (and band selection) is covered later in the book. For now, it is important to understand that an increase in output power of 3 dB doubles the output power.

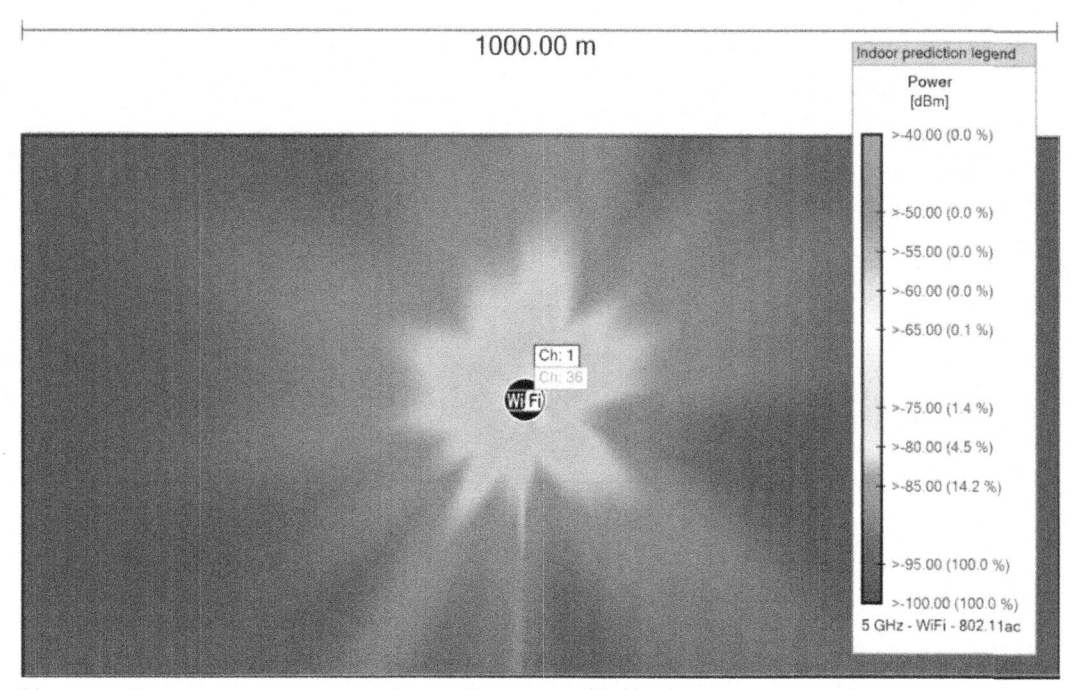

Figure 1.11: Attenuation at 5 GHz Illustrated in Free Space – image Generated by iBwave Wi-Fi

Understanding attenuation is a critical factor in selecting and recommending the appropriate equipment and configuration settings for that equipment. As a CWS, you do not have to understand the details of attenuation, but it is important to know that 5 GHz RF signals attenuate roughly twice as fast (because the wave travels in free space) as 2.4 GHz signals.

Absorption
The next phenomenon of which the CWS candidate should be aware is absorption. *Absorption* occurs as the RF wave passes through a medium that cannot respond to

the speed of the RF wave effectively on a molecular level; the result is a conversion of some of the RF energy to heat. Because of this energy loss, the RF wave exits the material with a significantly lower amplitude than it would have had if it were traveling through free space. This behavior must be considered when planning for WLANs and selecting and configuring equipment.

Often, engineers take wall attenuation measurements to understand the absorption levels of materials. This task is accomplished by placing an access point (AP) a few feet from the wall and then measuring the signal strength a few feet away on the other side of the wall. For example, if you place the AP three feet from a four-inch thick wall and measure the signal strength three feet from the wall on the other side, the difference in signal strength when compared to free space at the same distance shows a rough measurement of wall attenuation resulting from the absorption.

Most WLAN design tools allow a user to draw walls and other materials and then configure the attenuation factor for those walls. Many tools come with a library of standard materials and their attenuation levels. For example, you can choose concrete and even the type of concrete (e.g., with or without metal rebar). Drywall is typically included as well as other standard materials. These drawings and wall/floor specifications are created to simulate an environment in the design application. In many cases, engineers must go on-site and capture wall attenuation factors (measuring the loss incurred as RF waves travel through the walls) to implement a proper design document in a tool like iBwave Wi-Fi or Ekahau Site Survey, among others. The existing internal attenuation factors for the various building materials are approximations. Only on-site measurements can reveal exact values.

Impact on Signal Strength

With an understanding of reflection, attenuation, and absorption, you can now consider the impact on signal strength at the receiver. The signal strength is one of three key factors that impacts the receiver's ability to achieve higher data rates and, therefore, higher throughput on the network.

> *Data rate* is a phrase used to denote the speed at which bits are sent on an RF medium at the lowest layer of networking (called the Physical Layer or PHY). *Throughput* is the term used to denote higher-layer data that is useful for applications and user actions. The throughput is the effective rate at which data is transmitted or received by the user. Throughput is always lower than the data rate of the Wi-Fi link.

Understanding the factors that impact data rates is important, even when selecting equipment. Equipment components (e.g., radios, antennas) affect the way signals propagate from devices. Therefore, it is important for the CWS to understand the factors that impact data rates so that effective decisions can be made.

Three primary factors impact data rates:

- **Signal Strength:** The amplitude of the received signal.

- **Noise Floor:** The general RF noise (RF waves other than the desired signal) in the environment, which is also called ambient noise.

- **Interference:** Intermittent or consistent RF energy at amplitude levels that are sufficient to impact signal quality.

When only considering the signal strength and noise floor, a metric called the *signal-to-noise ratio* (*SNR*) can be evaluated. Understanding the SNR is the first step to achieving higher data rates. When you have more SNR (for example, 30 instead of 10), you can use more complex modulation types, and therefore, you can achieve higher data rates. Do not worry about what the actual SNR values mean yet. The meaning of the values are covered in the next few pages. For now, know that higher numbers are better when evaluating SNR. The SNR is a simple metric that does not consider other RF generators; instead, it only looks at the signal strength and the noise floor.

When considering the signal strength, noise floor, and interference, it is called the *signal-to-interference plus noise ratio* (*SINR*). The interfering RF generator may be far enough from the receiver, or it may emit low-amplitude waves, resulting in a reduced SINR at the receiver, but it may still allow reception at lower data rates. The point is simple: whether considering SNR or SINR, more space between the desired signal and the other RF energy at the location of the receiver results in higher data rates.

> Interference occurs at the receiver and not between the transmitter and the receiver. RF waves "pass through" other RF waves continuing in the direction of propagation. The RF activity between endpoints is not as important as the activity at the endpoints.

So, how do we measure and represent the signal strength and the RF noise in the environment? You should understand the following primary metrics:

- Watt and milliwatt
- dB
- dBm
- RSSI

Watt and Milliwatt

The actual output power at transmitters is defined in watts and *milliwatts*. While outdoor wireless bridge links may use 1 to 4 watts (W) of output power, most indoor access points (APs) use much lower output power levels measured in milliwatts. A milliwatt is 1/1000 of a watt. Therefore, 100 milliwatts (mW) is 1/10 of a watt. Configuration interfaces may show the output power in mW, dBm (which is covered next) or using a simple number to represent the output power level. When a simple number is used, you must explore the vendor documentation to determine the actual output power for each value. The W and mW are absolute power measurements. They are not relative to some other power level but are specific power levels themselves.

> An *Access Point (AP)* is a device that provides wireless connectivity to other devices (clients) in a defined area. The AP includes one or more wireless radios and one or more wired network connections

> (typically Ethernet). It allows wireless clients to connect to the wired network and vice versa.

The point to remember is that watts and milliwatts are used to indicate the output power at the transmitter, but they are not used to measure the received power at the receiver. Received power levels are usually far too low to be represented in watts or even milliwatts in a meaningful way. Therefore, another metric must be used: dBm. To understand dBm, you must first understand the decibel, which the next section explores.

dB and dBm

The *decibel* (*dB*) is a relative measurement value—that is, it measures the difference between two power levels. For example, it is common to say that a particular power level is 6 dB stronger than another power level or that it is 3 dB weaker. These statements mean that a 6 dB gain and a 3 dB loss has occurred, respectively.

A decibel is 1/10th of a bel. You could equally say that a bel is 10 decibels. The decibel is based on the bel, which was developed by Bell Laboratories to calculate the power losses in telephone communications as ratios. The definition of a bel is simple: 1 bel is a ratio of 10:1 between two power levels. Therefore, a power ratio of 200:20 is 1 bel (10:1), 200:40 is .5 bel (5:1) and 200:10 is 2 bel (20:1). In the end, the decibel (dB) is a measurement of power that is frequently used in RF mathematics.

dBm is an absolute measurement of power, where the *m* stands for milliwatts. Effectively, dBm references decibels relative to 1 milliwatt, that is, 0 dBm equals 1 milliwatt. Once you establish that 0 dBm equals 1 milliwatt, you can reference any power strength in dBm. Because a wireless receiver can detect and process extremely weak signals, it is easier to refer to the received signal strength in dBm rather than in mW. For example, a signal that is transmitted at 4 W of output power (4000 mW or 36 dBm) and experiences -63 dB of loss has a signal strength of .002 mW (-27 dBm). Rather than saying that the signal strength is .002 mW, we say that the signal strength is -27 dBm.

The best tools denote received signal strength in units of dBm. dBm has become the standard for representing received signal strength. Interestingly, some tools state that they report Received Signal Strength Indicator (RSSI) values, but they report dBm values. You will understand this more when you explore RSSI next. While most users rely on bars (e.g., three bars of signal, four bars of signal), percentages or some arbitrary value, the WLAN specialist depends on greater accuracy, which is provided by the dBm value.

RSSI

The *Received Signal Strength Indicator (RSSI)* is a variable or arbitrary metric defined in the 802.11 standard. The IEEE 802.11 rules that exist for RSSI implementation state that it is optional, it should report the rating to the device driver, and it should use 1 byte for the rating and exhibit a potential range of 0 to 255. Vendors choose a range within the IEEE-defined range. For example, a vendor may choose a range from 0 to 100 or from 0 to 60. Since the range is used to express the received signal strength, an RSSI value of, say, 30 is meaningless without knowing its range. An RSSI of 30 with a range of 0-60 is roughly 50% of the signal strength, but the same value of 30 with a range of 0-100 is roughly 30% of the signal strength.

> The 802.11 standard is explored in more detail in Chapter 3 (*The 802.11 Standard*). In that chapter, the process of creating a standard as well as a summary of the 802.11 standard itself is provided.

Due to potential difficulties associated with the interpretation of the meaning of RSSI, dBm has effectively replaced RSSI as the metric that is used by most WLAN engineers to denote true received signal strength. However, it is important to remember that some WLAN scanning tools will report dBm signal strength with a label of RSSI. The WLAN specialist should read the tool's documentation to learn whether RSSI is reported or if dBm is reported with an RSSI label. More consistency and accuracy is needed in this area concerning tool development and terminology.

Signal Strength in WLAN Scanners

To illustrate the ability to measure and report signal strength, this section will show several tools that provide the information. The first tool is the NETSH command available from the Windows Command Prompt. NETSH provides the NETSH WLAN SHOW NETWORKS MODE=BSSID command that lists all detected wireless networks as well as the signal strength of those networks. However, the signal strength, as shown in Figure 1.12, is reported as a percentage, which is not helpful for obtaining an accurate measurement of the signal strength. Therefore, it is useful to understand the formula that is used to convert the NETSH percentages to dBm levels.

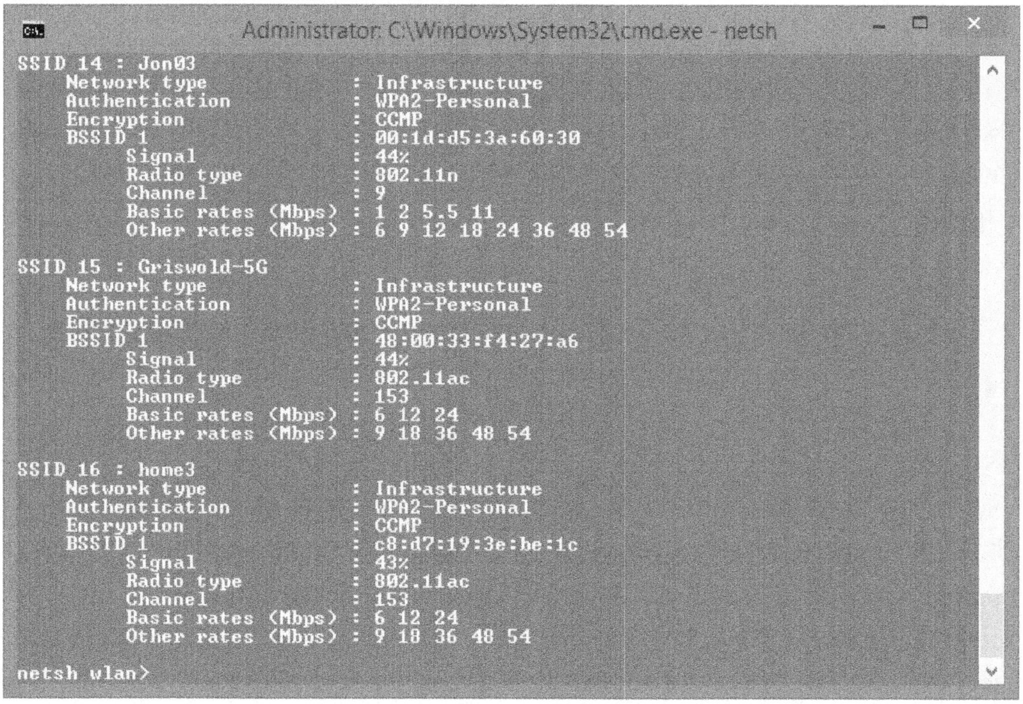

Figure 1.12: Signal Quality Shown in NETSH

The following algorithm explains this conversion:

```
If (signal <= 0%)
    dBm = -100
else if (signal >= 100%)
    dBm = -50 or better
else
    dBm = (signal / 2) - 100
```

Therefore, if the percentage is 40%, the dBm calculation is as follows:

```
dBm = (40 / 2) - 100
dBm = 20 - 100
dBm = -80
```

Similarly, if the percentage is 70%, the dBm calculation is as follows:

```
dBm = (70 / 2) - 100
dBm = 35 - 100
dBm = -65
```

Graphical tools are also useful, and a modern example is Acrylic Wi-Fi. Tools like this show the detected networks, their signal strength at the current location and other details such as channel, supported connection types, maximum data rates and security settings. Figure 1.13 shows Acrylic Wi-Fi Home. A professional edition is also available that contains added features. As discussed previously, this tool states that it reports RSSI values, but it actually reports dBm values. Additionally, Acrylic Wi-Fi, as well as other similar tools, uses a line graph to show signal strength over time. This feature can be very useful to detect variations in an environment. Even the most stable environments have some variations in signal strength (typically in the 3 to 10 dB range), but significant variations (variations higher than 5 or 6 dB) without client movement can indicate problems.

A similar tool that is available via macOS is Wi-Fi Explorer. Like Acrylic Wi-Fi, Wi-Fi Explorer shows the networks in the area as well as similar details to what

Acrylic Wi-Fi shows. It also shows signal strength over time and provides protocol decoding (viewing the actual 802.11 network communications), which is in the professional version of Acrylic Wi-Fi as well. Figure 1.14 shows the Wi-Fi Explorer tool.

Tools such as these allow you to determine the signal strength at a given location quickly. They are not as useful for reporting the signal strength at multiple locations without a great deal of extra effort on the part of the WLAN specialist. For such reporting, the use of actual site-survey or WLAN design software is much better. Software like Ekahau Site Survey, iBwave Wi-Fi, AirMagnet Survey Pro and TamoSoft Survey work well in these scenarios.

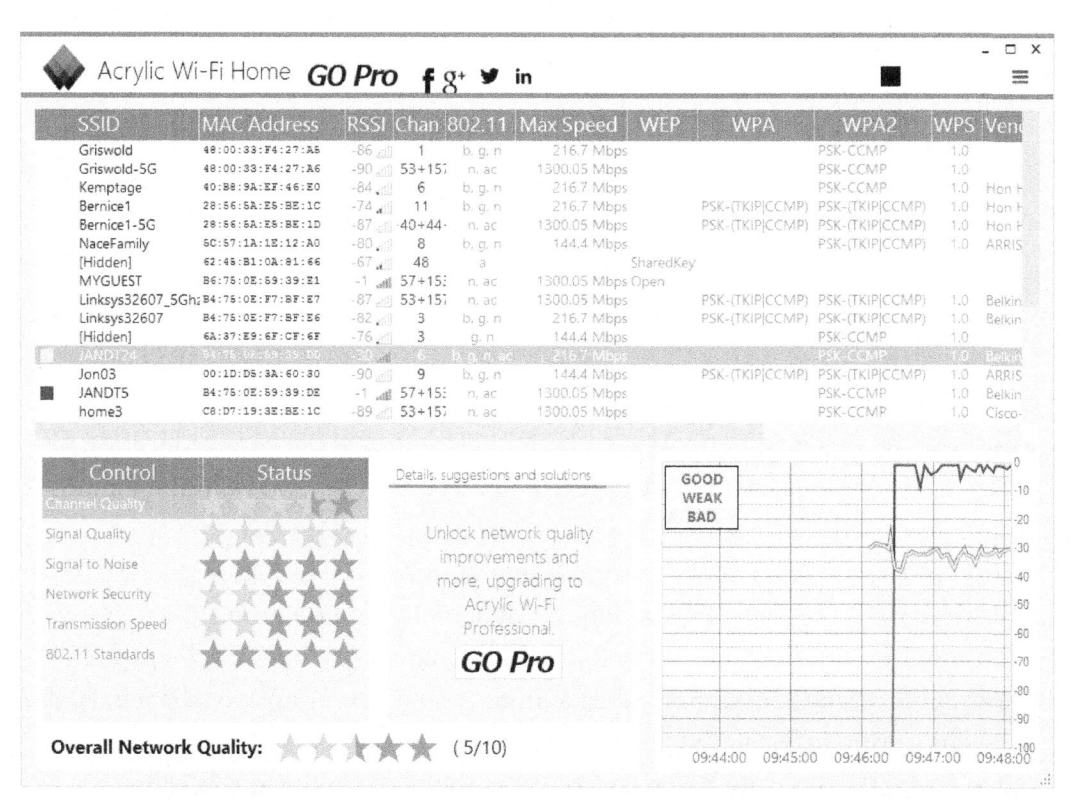

Figure 1.13: Acrylic Wi-Fi Home Showing Signal Strength

Such site survey and WLAN design tools allow you to import a floor plan; then, you can walk throughout the facility, click where you are located on the map, and move to the next location and repeat. The result will be gathered signal strength metrics throughout the facility. These tools can generate heat maps of the detected signal strength (among other metrics) and are very useful for post-installation analysis or troubleshooting scenarios where the WLAN is not performing as expected. Additionally, they can simulate environments for pre-sales reports and proposals, and they can be beneficial for problem remediation in existing networks.

Site survey details are beyond the scope of this book; however, as a CWS, it is important to know that site surveys and proper WLAN design are essential for a well-performing WLAN that meets the requirements of any deployment. In every case, the CWS should recommend that the customer or organization perform at least a minimal site survey (locating interferers and verifying attenuation factors such as walls and floors in the environment) and use expert designers to plan the WLAN. The significantly improved results are worth the extra effort and cost.

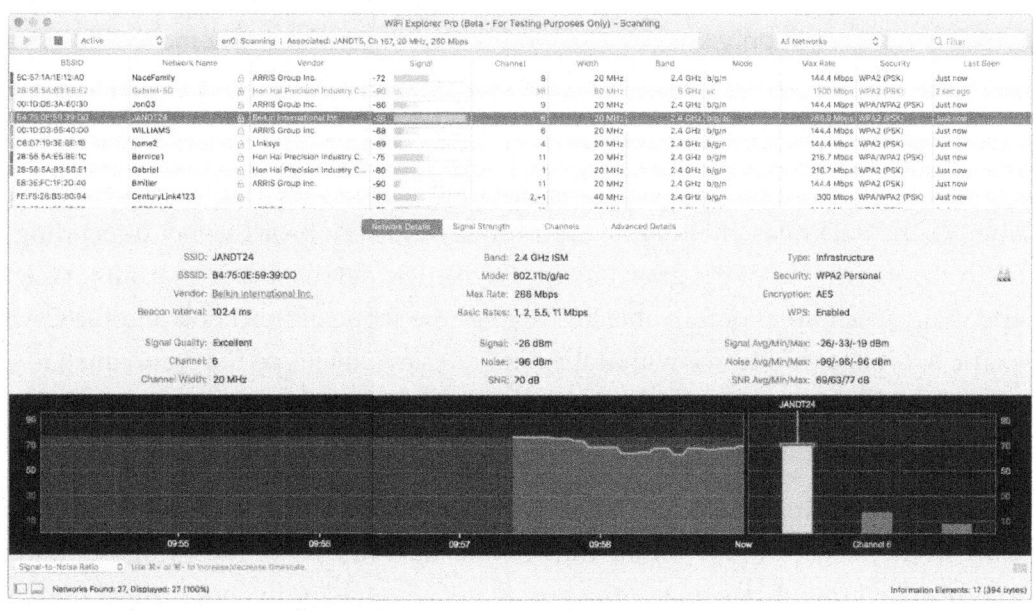

Figure 1.14: Wi-Fi Explorer on macOS

How Wi-Fi Uses RF

Now that you understand the basics of RF and the main RF properties, you can explore how Wi-Fi uses RF to transmit and receive data. The process utilizes modulation and coding.

Modulation is used to impress information onto RF waves. An RF wave that is used to carry data is called a carrier wave. When the wave is modulated, it is called a signal. So, when you hear the term signal, it refers to a modulated RF wave in the context of WLANs.

Coding is used to encode the data before modulation onto the RF carrier wave. The coding provides a method for some level of recoverability. That is, if the receiver is unable to receive the entire modulated wave stream correctly, it may be able to recover the original data because of the coding used. The modulation and coding together result in the data rate achieved. Note that there are also some additional elements that are important, but they are beyond the scope of this book or the CWS exam. When taken together, these elements combine to achieve a given data rate, which is now commonly known as a modulation and coding scheme (MCS). The 802.11 standard provides tables that represent the different MCS implementations available. Figure 1.15 shows one of these tables for an 802.11ac device.

In Figure 1.15, you can see that, by using a modulation called BPSK and a coding rate (R), the data rates of either 27 Mbps or 30 Mbps can be achieved, depending on the configuration of the guard interval (GI). The details of the guard interval and many other areas of the table are beyond the scope of this book and the CWS exam, but it is useful to know that the data rates available are fixed and are based on tables like this. It is important to know that this information is found in the 802.11 standard.

If the number of spatial streams changes, or the channel width changes, a different table would be used to determine the possible data rates. The most important thing to take away from this is that data rates do not change fluidly, but instead they change specifically. Data rates decrease or increase to specific lower or higher data rates, respectively. They do not drop from 60 Mbps to 59 Mbps, for example;

instead, as seen in Figure 1.15, they drop from 60 Mbps to 30 Mbps or increase from 60 Mbps to 90 Mbps.

The modulations used in 802.11 devices include the following:

- Binary Phase Shift Keying (BPSK)
- Quadrature Phase Shift Keying (QPSK)
- Quadrature Amplitude Modulation (QAM)

VHT-MCS Index	Modulation	R	N_{BPSCS}	N_{SD}	N_{SP}	N_{CBPS}	N_{DBPS}	N_{ES}	Data rate (Mb/s) 800 ns GI	Data rate (Mb/s) 400 ns GI
0	BPSK	1/2	1	108	6	216	108	1	27.0	30.0
1	QPSK	1/2	2	108	6	432	216	1	54.0	60.0
2	QPSK	3/4	2	108	6	432	324	1	81.0	90.0
3	16-QAM	1/2	4	108	6	864	432	1	108.0	120.0
4	16-QAM	3/4	4	108	6	864	648	1	162.0	180.0
5	64-QAM	2/3	6	108	6	1296	864	1	216.0	240.0
6	64-QAM	3/4	6	108	6	1296	972	1	243.0	270.0
7	64-QAM	5/6	6	108	6	1296	1080	1	270.0	300.0
8	256-QAM	3/4	8	108	6	1728	1296	1	324.0	360.0
9	256-QAM	5/6	8	108	6	1728	1440	1	360.0	400.0

Figure 1.15: MCS Table for 802.11ac 2 Stream Devices with a 40 MHz Channel

BPSK and QPSK only use phase shifts (changes in the phase of sequential waves) for modulation. BPSK uses two phase shifts, and QPSK uses four. This difference is why the data rate doubles from MCS index 0 to MCS index 1 in Figure 1.15. QAM uses both phase shifts and amplitude shifts (changes in the signal strength of sequential waves), allowing for even higher data rates. The numbers before each QAM modulation name indicate the total number of waveforms possible with the modulation. For example, 16-QAM uses 16 different waveforms, and 64-QAM uses 64 different waveforms. Note that a waveform is some combination of a phase and amplitude.

> The specific details of how the modulation works is beyond the scope of this book and the CWS exam, but it is useful to understand the terminology and know that different modulation and coding combinations within the same channel width and number of spatial streams results in different data rates. Higher data rate modulations use more complex waveforms that can represent more data in the signal.

The result of this modulation and coding information is that the highest data rate that is achieved with a two-stream device operating on a 40 MHz channel is 400 Mbps.

Table 1.1 lists the highest data rates that are achievable for different combinations of spatial streams and channel widths. Only 20-MHz and 40-MHz channels are listed because these are the most practical channels to use in enterprise WLAN deployments. While 802.11ac supports 80-MHz and 160-MHz channels, enterprise engineers rarely use 80-MHz channels, and 160-MHz channels should not be used in any enterprise WLAN deployment unless regulatory agencies provide a considerable amount of additional frequency space at some point in the future.

Channel Width	Spatial Streams	Maximum Data Rate (Mbps)
20 MHz	1	86.7
20 MHz	2	173.3
20 MHz	3	288.9
20 MHz	4	346.7
40 MHz	1	200
40 MHz	2	400
40 MHz	3	600
40 MHz	4	800

Table 1.1: Maximum Data Rates for 802.11ac Devices with 20 MHz and 40 MHz Channels

More details on this issue are presented in Chapter 4 (*Channel Plans and Performance Factors*). Ultimately, the channel width, number of spatial streams, and quality of the received signal are the main factors that impact data rates and, therefore, performance on a WLAN. If these factors are chosen appropriately, the WLAN will perform much better; if they are not, no amount of configuration tweaking is likely to result in the required performance. The best practice is always to use a well-trained WLAN engineer to design any WLAN. Merely installing APs and hoping for the best is never a good strategy.

Chapter Summary

In this chapter, you learned about the foundation of all WLANs: radio frequency (RF). In the process, you learned about RF waves and wave characteristics, including wavelength, frequency, amplitude and phase. Next, you explored RF behaviors, including reflection, attenuation and absorption. Finally, you learned about signal strength measurements and modulation and coding schemes used to make WLANs work and provide effective communications. In the next chapter, you will learn about RF hardware and the importance of selecting the right hardware for WLAN deployment.

Points to Remember

Remember the following important points:

- Radio frequency (RF) waves are used to transmit and receive data on WLANs.

- The wavelength of a wave is related to its frequency: higher frequencies correspond with shorter wavelengths, and lower frequencies correspond with longer wavelengths.

- The amplitude of a wave is the power or strength of the wave.

- The phase of a wave is a comparative attribute that relates one wave to another.

- Waves may be 'in phase' with other waves or, to some degree, out of phase with other waves.

- Modulation uses phase shifts (PSK) and amplitude shifts (ASK) in WLANs.

- RF waves are reflected off obstacles that are large in relation to their wavelengths and that have reflective properties.

- Many materials absorb some of the RF energy of the waves as the waves pass through them.

- RF transmission power is often referenced in watts (W) or milliwatts (mW), but received signal strength is typically referenced in dBm.

- 802.11 devices use a modulation and coding scheme (MCS) to achieve the data rate in the link.
- In addition to the modulation and coding, the channel width and number of spatial streams impact the achievable data rate.

Review Questions

1. What denotes the strength of an RF wave or signal?
   a. Wavelength
   b. Phase
   c. Amplitude
   d. Modulation

2. What is decreased as the frequency of a wave increases?
   a. Wavelength
   b. Phase
   c. Modulation
   d. Coding

3. What RF behavior is similar to light waves as seen in a mirror?
   a. Absorption
   b. Attenuation
   c. Reflection
   d. Phase

4. What kind of modulation is not used in WLANs?
   a. PSK
   b. FSK
   c. ASK
   d. BPSK

5. What is the name of the scheme that determines data rates in WLANs?
   a. Modulation and Coding Scheme
   b. Reflection and Attenuation Scheme
   c. Moore's Law
   d. Modulation and Phase Scheme

6. What happens when RF energy is converted to heat when passing through some materials?
   a. Absorption
   b. Reflection
   c. Phase shifting
   d. Frequency shifting

7. What modulation type uses only Phase Shift Keying?
    a. BPSK
    b. 16-QAM
    c. 64-QAM
    d. 256 QAM

8. In addition to modulation and coding, what other two primary factors impact the data rate that is achievable in a WLAN link?
    a. Frequency and phase
    b. Channel width and number of spatial streams
    c. Channel width and frequency band
    d. Frequency band and spatial streams

9. What comparative attribute relates an RF wave to another wave in the same channel?
    a. Wavelength
    b. Frequency
    c. Amplitude
    d. Phase

10. What metric is preferred by engineers and is used to represent received signal strength?
    a. RSSI
    b. dB
    c. dBm
    d. Frequency

Review Answers

1. **C is correct.** The amplitude is the strength or power of the RF wave.

2. **A is correct.** As the frequency increases, the wavelength decreases. Alternatively, as the wavelength increases, the frequency decreases.

3. **C is correct.** Reflection occurs in RF waves as well as light waves and is analogous to light in a mirror.

4. **B is correct.** Frequency Shift Keying (FSK) is not used in WLANs. Both ASK and PSK are used, and BPSK is a form of PSK modulation.

5. **A is correct.** The Modulation and Coding Scheme (MCS) determines the data rates that are available. It is a factor of modulation, coding, channel width and spatial streams, among other variables.

6. **A is correct.** Absorption occurs when RF energy is converted to heat, and the result is an attenuation of the RF signal as it exits the absorbing medium.

7. **A is correct.** All Quadrature Amplitude Modulation (QAM) types use both ASK and PSK. BPSK and QPSK are the two modulation techniques that only use PSK.

8. **B is correct.** The channel width impacts the data rate, and wider channel widths support higher data rates. The number of spatial streams also impacts the data rate. More spatial streams with the same or better signal quality results in higher data rates.

9. **D is correct.** The phase of an RF wave is an attribute that compares one wave with another wave. They can either be 'in phase' or, to some degree, out of phase.

10. **C is correct.** While RSSI is a common metric as well, engineers prefer dBm because it is an absolute power measurement, whereas RSSI is arbitrary and uniquely specified by WLAN vendors.

Chapter 2 – RF Hardware

Objectives Covered
1.3 Understand antenna types
      1.3.1   Omnidirectional
      1.3.2   Semi-directional
      1.3.3   Highly directional
      1.3.4   Internal vs. external

The primary RF hardware component that is covered in the CWS exam, other than APs and clients, is antennas. Antennas allow for the radiation and reception of RF signals. This chapter explains antennas and the various types available. Internal and external antennas are explored. Finally, additional useful RF hardware is briefly introduced.

Understanding Antennas

Antennas (also called aerials, though some contend an aerial is different) are radiating elements. This means that they radiate or release electromagnetic energy into the air. They are also receiving elements, meaning they receive (capture or pick up) electromagnetic energy from the environment. Because different frequencies have different wavelengths, antennas are engineered for best use with specific frequency ranges. For example, you may acquire an antenna that can be used for 2.4 GHz signals and another for 5 GHz signals. Antennas can be engineered to work for both frequency bands, but they will not work as optimally as those engineered for a specific band or range of frequencies.

The engineering of the antenna determines its radiation pattern. When selecting antennas, it is essential to choose the antenna that provides the desired radiation pattern. Today, most APs use internal omnidirectional antennas that radiate in a 3D circular pattern around the antenna (often said to be shaped like a donut). In many cases, such APs cannot connect to external antennas, and the WLAN design must accommodate for the internal antenna radiation pattern in deployment.

Many outdoor APs use external antenna connectors. The attached antennas can be very different in the coverage provided depending on the radiation pattern. So, in general, internal (indoor) APs use internal antennas, and external (outdoor) APs use external antennas, but exceptions do exist.

> The *radiation pattern* is the way in which the antenna radiates energy into the air. Some antennas radiate energy in a reasonably uniform circumference around them, and others focus most of the energy in a particular direction.

In the previous chapter, you learned the basics of RF waves. Some additional factors must now be considered. RF waves are electromagnetic waves. This statement means that they are comprised of electric fields and magnetic fields. The electric field gives rise to a magnetic field, and the magnetic field, in turn, gives rise to an electric field. This perpetual action is what causes RF waves to propagate through space.

Consider Figure 2.1 to understand this concept of electromagnetic propagation better. The E in Figure 2.1 represents the electric field, and the B represents the magnetic field. Also, Z represents time, and the Greek lambda (λ) represents the wavelength. The key factor here is to understand that an antenna that is impressed with electromagnetic energy will release that energy into free space with the electric field moving out parallel to the antenna element and the magnetic field moving perpendicular to it. In the opposite way, a receiving antenna can capture this energy.

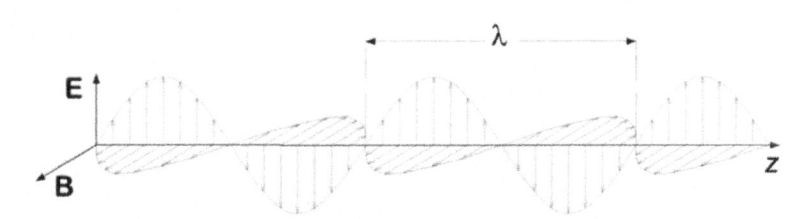

Figure 2.1: Electric and Magnetic Fields Causing RF Wave Propagation

RF Radiation Patterns
Varying antennas will radiate differently based on both their designs and the enclosures or reflective materials surrounding them. For example, an antenna that radiates equally in all directions around the radiating element but that is placed in front of a highly reflective concave surface will actually result in energy radiated mostly in one direction. This design is partly why highly directional antennas work as they do.

The iBwave Wi-Fi tool shows 3D radiation patterns for both external antennas and internal antennas in APs. These 3D patterns are based on antenna charts that will be discussed later in this section. For now, notice the very different radiation patterns from varying antennas and APs, as shown in Figures 2.2 through 2.5.

Figure 2.2 shows the typical radiation pattern of a patch or panel antenna. Such antennas are designed to radiate RF waves out from the antenna in one direction, often covering a 180-degree area. They are instrumental when you need to cover an area but do not want to cover the area behind the antenna. For example, when antennas are placed on walls and face inward toward the coverage area, they result in small signal losses outside the desired coverage area.

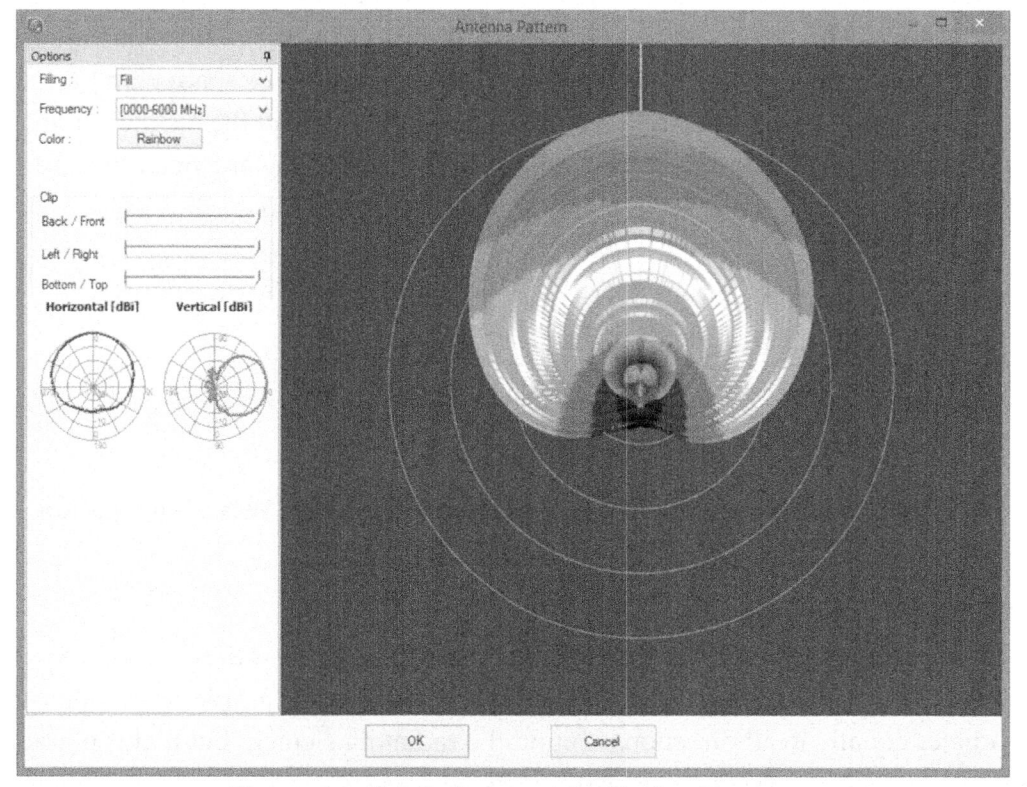

Figure 2.2: Patch Antenna Radiation Pattern

Figure 2.3 shows the typical radiation pattern of an omnidirectional antenna. Note that the omnidirectional antenna is the most common internal antenna type in APs, though internal antennas often lack the nearly perfect omnidirectional radiation pattern seen in Figure 2.3. Such antennas should be placed closer to the center of the desired coverage area.

Figure 2.4 shows the radiation pattern of an Aruba Networks AP. This model happens to be the AP-325, and the radiation pattern is generated by iBwave Wi-Fi. Using its internal antennas, you can see that the radiation pattern does not match the nearly perfect omnidirectional pattern experienced by a single dedicated external antenna, but this varied pattern is in part due to the mechanical makeup of the AP and the complexity of designing tiny internal antennas.

Figure 2.5 shows the Fortinet FortiAP 832i radiation pattern as represented in iBwave Wi-Fi. This AP shows some characteristics of a patch or panel antenna, though with a broader propagation path and omnidirectional in nature. When selecting and deploying APs, the CWS must understand the nature of antenna propagation expected so that they can be deployed in the appropriate locations.

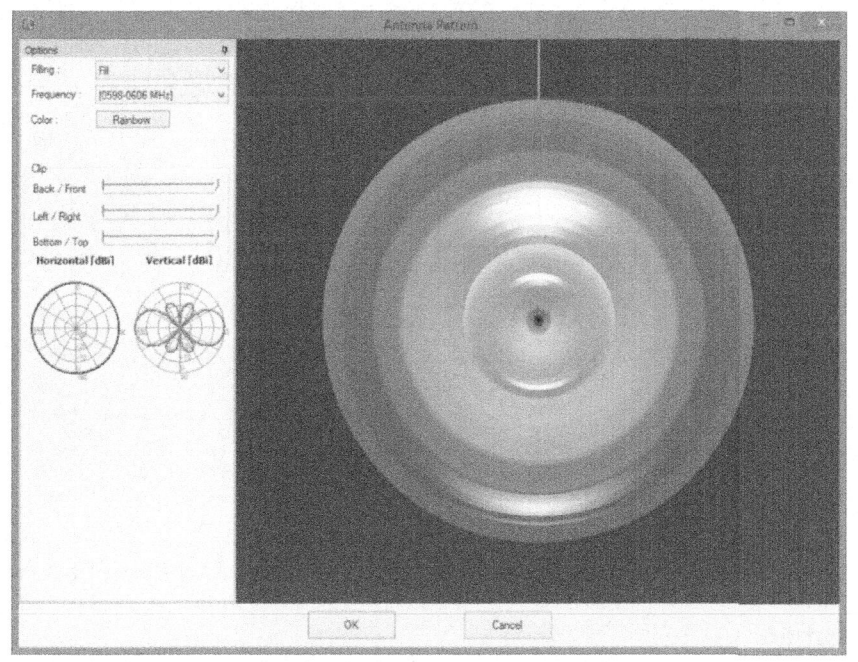

Figure 2.3: Omni Antenna Radiation Pattern

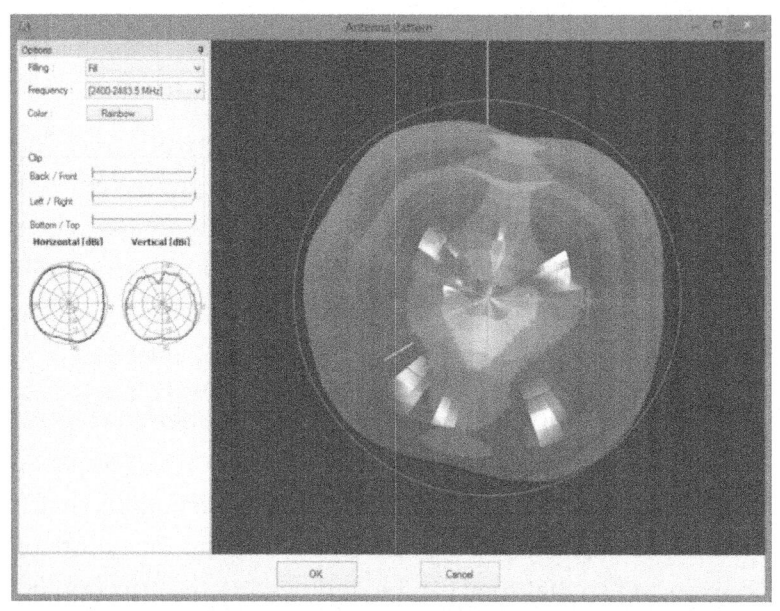

Figure 2.4: Aruba Networks AP-325 Access Point Radiation Pattern

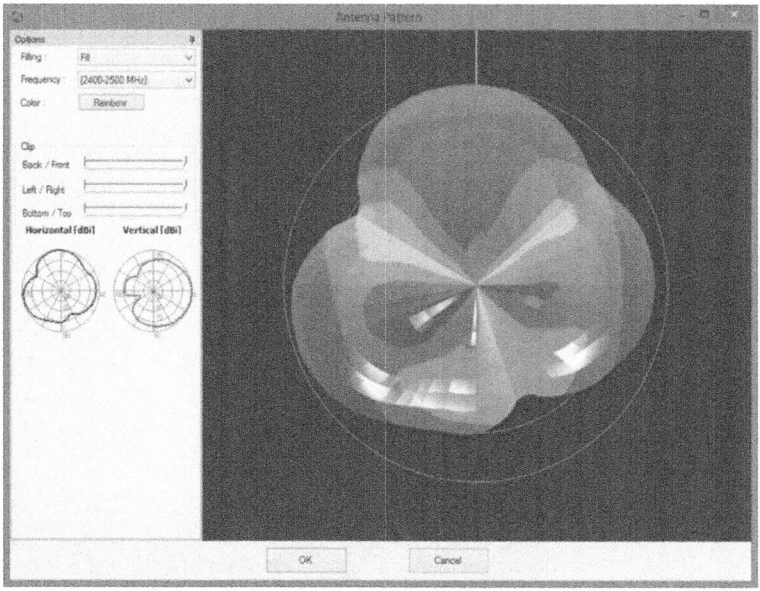

Figure 2.5: Fortinet FortiAP 832i Access Point Radiation Pattern

RF Reception

RF reception occurs when the electromagnetic wave passes by an antenna, and the antenna picks up the energy passing it down the antenna, through connectors and possibly cables to the radio receiver. An antenna receives best in the same direction that it propagates. For example, the patch antenna pictured in Figure 2.2 would receive signals best that enter on the same side of the patch antenna as the signals that radiate outward. Stated differently, an antenna hears best where it talks best.

> Filters are often used in addition to a radio, but for the purposes of this discussion, it is sufficient to know that the energy is passed from the receiving antenna through some path of connectors and/or cables to the radio.

In the general field of wireless communications, a concept exists that says, "If you can hear me, I can hear you." This saying means that, if the output power level on both ends of a link is similar, the antenna gain that allows for transmissions to reach the other end of the link will also allow for reception of signals from the other end of the link. That is, the antenna that provides an increase in directional RF energy to the transmitted signal also has the ability to receive relatively weaker signals from that same direction.

Active vs. Passive Gain

All antennas create RF gain. The differences among them are in the areas of directionality or focus as well as the amount of gain. The gain created by an antenna is called *passive gain*. No more energy exists than that which was input into the antenna, but the energy is radiated more in one direction than another, and so there is gain. Passive gain occurs due to the directionality of the antenna. *Active gain* requires an amplifier, which is a device that actually increases the RF energy before passing it through the antenna.

> Remember that antennas cause passive gain and amplifiers cause an active gain. Passive gain is achieved by focusing the RF energy more in one direction than another and an active gain is achieved by increasing the actual amplitude of the signal before injecting it into the antenna.

dBi

The gain of an antenna is measured in *dBi* (decibels compared to an isotropic radiator). Another measurement called dBd was used for WLAN antennas early on but is rarely used today. dBd was defined as decibels compared to a dipole radiator, which had 2.14 dBi gain. In the end, comparing an antenna to some arbitrary standard of gain (i.e., 2.14 dBi) was not useful, so dBi has become the standard for antenna gain in the Wi-Fi industry.

An *isotropic radiator* is a zero-gain antenna. This definition means that it radiates energy equally around the radiation point in a spherical pattern. No such antenna exists, but it is the virtual reference point for dBi. That is, the difference between what an isotropic radiator would radiate if it existed, and the actual radiation of the antenna is the gain in the direction(s) of radiation. Common antenna gain values include 2, 3, 5, 7, and 11 dBi, among others.

> The details of RF math are reserved for the CWNA exam, but it is useful to know some rough values. 3 dB of gain doubles the power, and 3 dB of loss halves the power. 10 dB of gain increases the power by a factor of 10, and 10 dB of loss decreases the power by a factor of 10.

A 3-dBi antenna radiates twice the amount of power in the intended direction than an isotropic radiator would. To be clear, adding 3 dB of power doubles the power and subtracting 3 dB of power halves the power. A 10-dBi antenna radiates ten times the amount of power in the intended direction.

Antenna Charts

Most antenna vendors and many AP vendors provide charts that show the expected propagation pattern for their antennas. These charts help you to understand the expected coverage you can achieve. The pattern will remain the same, in free space, if the power is increased or decreased; it will simply cover a larger or smaller area respectively. However, for indoor networks, reflections and absorption will result in entirely different coverage patterns than expected based on the antenna chart.

Notice in Figure 2.6 that the row of filing cabinets and the internal thick and thin walls change the pattern of the radiating antennas from a pure omnidirectional pattern to a pattern that is entirely unique to that space. In this image, the antenna is on the left of the thick wall, and the AP (labeled as Wi-Fi) is on the right of the thick wall, and they are connected with a coaxial cable. Therefore, the antenna chart helps you to determine the general propagation pattern but not the propagation that will actually occur in an indoor environment.

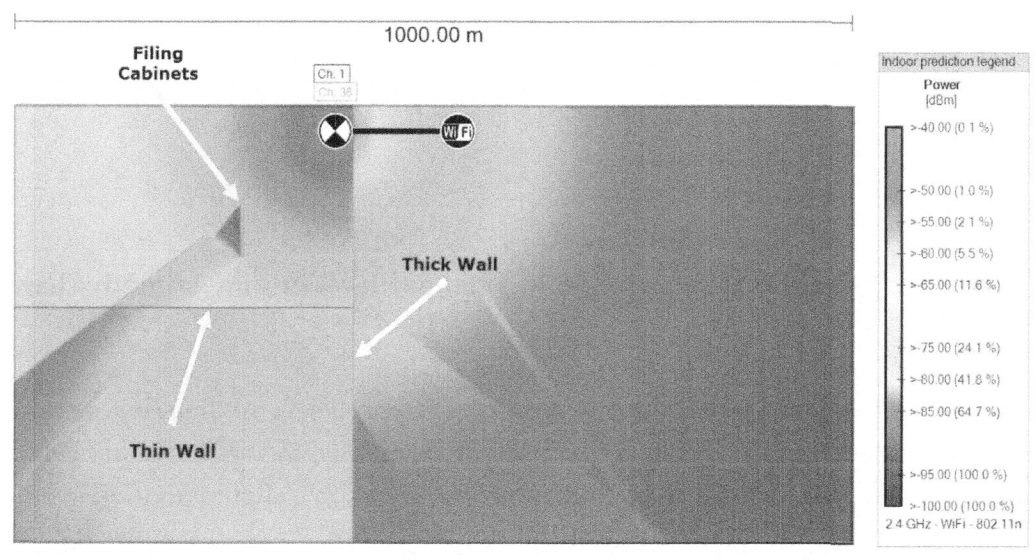

Figure 2.6: iBwave Representation of Propagation Patterns with Reflection and Internal Walls

Two basic antenna charts are the azimuth and elevation charts. The azimuth chart shows a view that you would see if you were to look down at a dipole antenna. The azimuth chart shows how the energy radiates out from the antenna in a top-down view based on intended use. The elevation chart shows a view that you would see if you look at the antenna from the side. It shows how the energy radiates out from the antenna in a side view based on intended use.

Figure 2.7 shows the azimuth (horizontal) and elevation (vertical) charts for an omnidirectional antenna.

Antenna Types

In this section, the three most common antenna types will be reviewed: omnidirectional, semi-directional and highly directional. For each antenna type, the characteristics of the antenna type will be explained, and images of actual antennas as well as matching antenna charts will be provided.

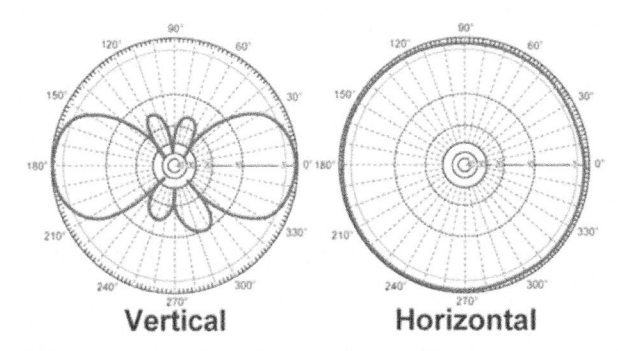

Figure 2.7: Omnidirectional (Dipole) Antenna Elevation and Azimuth Charts

Omnidirectional

Omnidirectional antennas, which have been referenced several times before now, radiate in a donut-like pattern in 3D space. Horizontally, they radiate out nearly uniformly in all directions around the antenna (with some slight variation in some antennas). Vertically, they reveal the donut-like pattern. Figure 2.8 shows the Cisco AIR-ANT4941 omnidirectional antenna.

Figure 2.8: Cisco AIR-ANT4941 2 dBi Omnidirectional Antenna

Figure 2.9 shows the elevation and azimuth charts for the Cisco AIR-ANT4941 antenna. The left part of the image shows the elevation or vertical chart (also called the E-plane). The right part of the image shows the azimuth or horizontal chart (also called the H-plane).

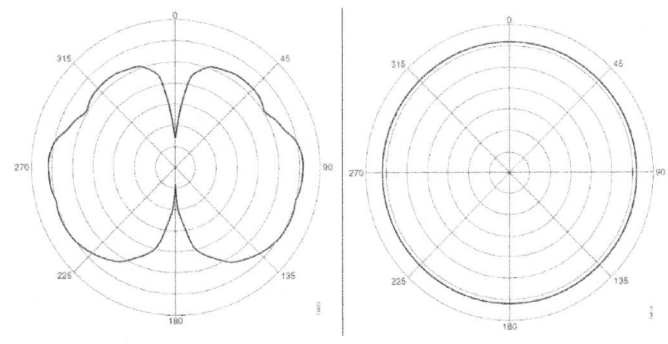

Figure 2.9: Cisco AIR-ANT4941 Elevation (Vertical) and Azimuth (Horizontal) Charts

Figure 2.10 shows the same antenna along with detailed specifications. Notice that it is designed to operate in the 2.4 GHz (2402-2495 MHz) frequency band. It has a 50-Ohm impedance, which you do not need to understand in detail for the CWS exam or to play the role of a CWS. However, most WLAN equipment uses 50-Ohm cables and connectors. All cables and connectors in the link, including the antenna, should be 50-Ohm rated. Finally, the connector type is RP-TNC, which is required to know in order to ensure it is of the same type used by the intended AP, cable or connector to which you plan to connect it.

Antenna type	Dipole
Operating frequency range	2402-2495 MHz
Nominal input impedance	50 Ohms
2:1 VSWR bandwidth	2385 - 2515 Mhz
Peak gain	2 dBi
Polarization	Linear, vertical
E-Plane 3-dB beamwidth	70 degrees
H-Plane 3-dB beamwidth	Omnidirectional
Dimensions	5.5 in. (13 cm)
Weight	1 oz.
Connector type	RP-TNC plug
Environment	Indoor
Operating temperature range	32°F to 140°F (0°C to 60°C)

Figure 2.10: Cisco AIR-ANT4941 Specifications

Figure 2.11 shows the elevation chart of the same antenna with an illustration of the antenna superimposed over it. With this image, you can clearly understand how the elevation chart works.

Omnidirectional antennas are used in or near the center of the intended coverage cell or area. They are the most common internal antenna type used within APs. When you see antennas sticking out of APs or consumer-grade wireless routers, they are nearly always omnidirectional antennas as well. Higher gain omnidirectional antennas create a flatter vertical (elevation chart) pattern and therefore push the RF energy further out. However, they also cover less area above and below the main lobe. Figure 2.12 shows a high-gain omnidirectional antenna on the left and a lower-gain antenna on the right, depicting this variation in vertical signal coverage.

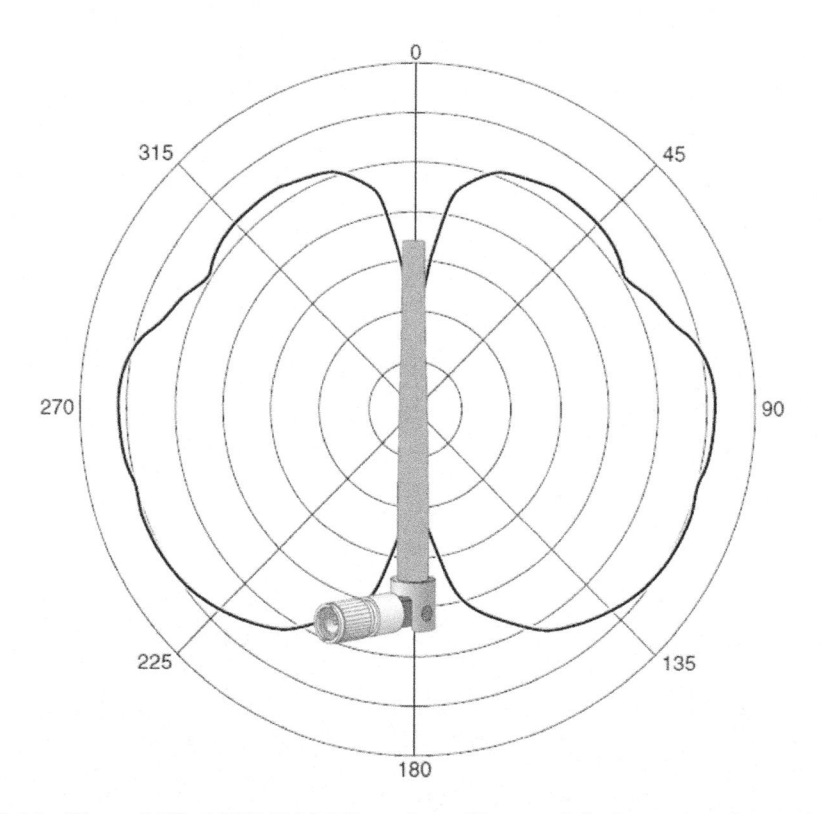

Figure 2.11: Cisco AIR-ANT4941 Elevation Chart with Antenna Superimposed

> The *main lobe* is the primary intended direction in which an antenna is designed to propagate. Some energy will always be propagated in other directions, but the main lobe is the direction of intention. For semi- and highly directional antennas, they also have rear lobes and side lobes as you will see in later charts within this chapter.

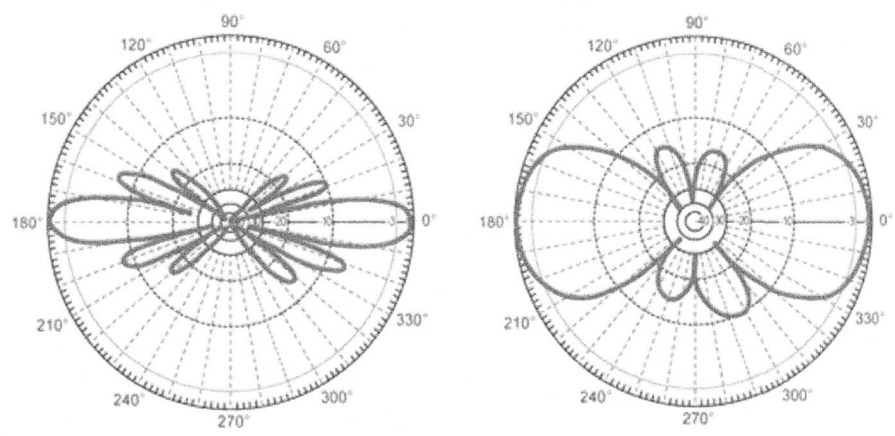

Figure 2.12: A High-Gain Omnidirectional Antenna (left) and a Normal-Gain Omnidirectional Antenna (right)

In many cases, higher-gain omnidirectional antennas are used when large outdoor areas with low user counts (called low density) must be covered. Additionally, they can be used in multi-floor facilities to help minimize signals from leaking through to the floors above and below. By mounting such antennas at medium heights and reducing the output power of each AP, cross-floor interference can be reduced. This is not the most common practice due to the cost of implementation, but it is a useful solution in some dense deployment scenarios.

Semi-Directional

Directional antennas can be subcategorized as either semi-directional or highly directional. Directional antennas are different from omnidirectional antennas in that they propagate the vast majority of RF energy in a horizontal pattern that is typically between 8 and 180 degrees. Semi-directional antennas usually cover about half the area of an omnidirectional antenna (usually between 50 and 180 degrees) or less. They are typically placed on the edge of the target coverage area and aimed inward. For example, they may be placed on a wall with the RF energy radiating in toward an office space. They can be instrumental in reducing co-channel interference (CCI) caused by APs that can hear each other.

> *CCI* occurs when multiple 802.11 devices are on the same channel but connected to different APs. All 802.11 devices in the same channel that can hear each other at some defined level must wait for each other's communications to end before transmitting.

Most semi-directional antennas used in WLANs are in either the patch or panel form. These antennas are usually in square-shaped containers and can be mounted on walls, poles, and even towers. Figure 2.13 shows a panel antenna from AccelTex Solutions, and Figure 2.14 shows the antenna charts for this antenna. Notice the coverage pattern created by the panel antenna. The number of elements in the antenna indicates the supported MIMO configurations or diversity configurations. MIMO uses multiple antennas for simultaneous transmission and reception. Diversity uses the best single antenna for transmission or reception.

Figure 2.13: AccelTex Solutions ATS-OP-245-47-6RPSP-36 Six Element Patch Antenna

The ATS-OP-245-47-6RPSP-36 antenna shown in Figure 2.13 is a six-element antenna. For example, it can be connected to a 3x3:3 MIMO AP that is dual-band. This means that the AP works at both 2.4 GHz and at 5 GHz. Three of the elements will be used for 2.4 GHz and three for 5 GHz.

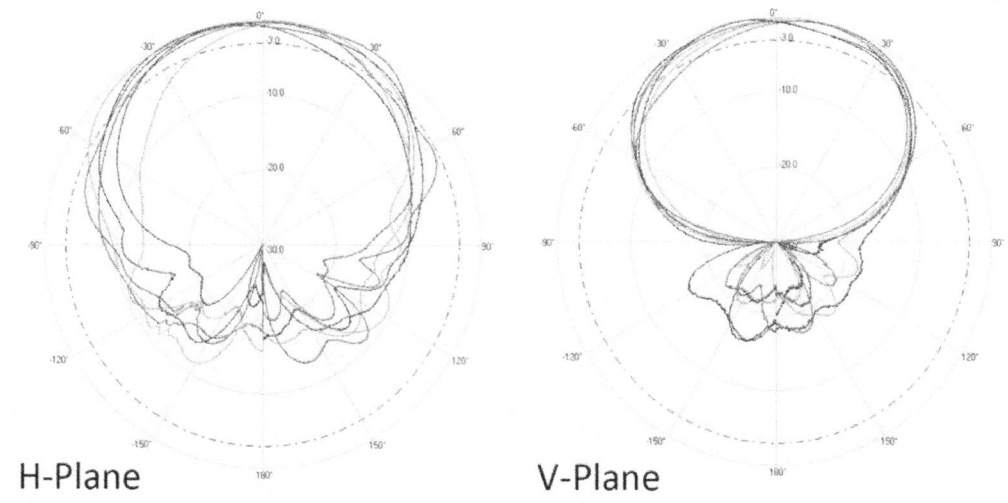

H-Plane   V-Plane

Figure 2.14: AccelTex Solutions ATS-OP-245-47-6RPSP-39 Azimuth (left) and Elevation (right) Charts

The nomenclature 3x3:3 (Tx x Rx : SS) indicates three transmitting (Tx) antennas, three receiving (Rx) antennas and three spatial streams (SS), meaning that all three antennas can be used to transmit or receive at the same time. Of course, you can transmit on all three or receive on all three, but you cannot both transmit and receive at the same time. The radio must be either "listening" or "talking," but not both simultaneously.

Notice in Figure 2.14 that the antenna charts use multiple colors. These colors represent the different elements of the antenna. Six elements exist, so six different colors are used to show the different radiation patterns of these elements.

Highly Directional
Highly directional antennas are usually used for bridge link connections that cover a significant distance (many meters or even kilometers). They are most often in the parabolic dish, grid or Yagi form factors. Highly directional antennas have a

radiation pattern of less than 50 degrees in the horizontal and vertical planes. In fact, 50 degrees would be considered a maximum pattern for many engineers to qualify them as highly directional instead of semi-directional.

> A *bridge link* is a connection between two separate networks or locations that allows data to be transmitted between them. Bridge links are often used to connect separate buildings so that the devices in one can communicate with the devices in the other and vice versa.

Because of the high directionality, these antennas can often have passive gain levels between 12 and 30 dBi and in some rare cases even more. When you consider the typical 24 dBi parabolic dish antenna with just 100 mW of input power, the radiated power (technically called the Equivalent Isotropically Radiated Power (EIRP)) would be 25.6 watts. This is a significant amount of power and can allow for links over great distances.

> *EIRP* is the output power of the radiating system at the antenna, including all gains and losses leading up to the antenna and the passive gain generated by the antenna itself. Most regulatory domains restrict the amount of EIRP, and many also restrict the amount of power inserted into the antenna.

Figure 2.15 shows a Yagi antenna. Yagi antennas are designed with multiple director elements that capture the RF energy and are often fully enclosed in a radome cover to protect the director elements and antenna from weather damage. Yagi antennas commonly provide gain levels that range from 8 to 20 dBi.

Figure 2.16 shows a parabolic dish antenna. These antennas use a reflector backing to enhance the directionality. They can have gain levels higher than 30 dBi.

Figure 2.17 shows the marketing information for a grid antenna. Like the parabolic dish antenna, they offer high gain, but they also have openings in the back panel to allow for better wind loading.

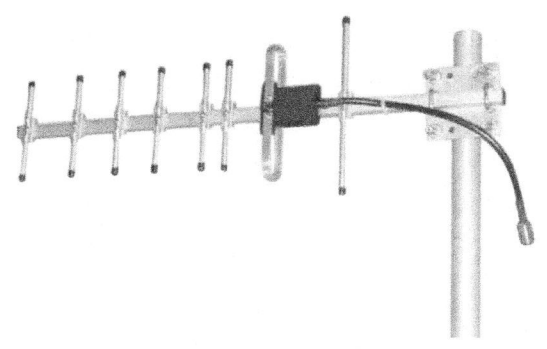

Figure 2.15: Yagi Antenna

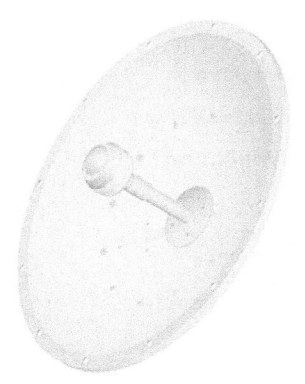

Figure 2.16: Parabolic Dish Antenna from Ubiquiti Networks

Both the parabolic dish and grid antennas are commonly used in long distance bridge links using WLAN APs or bridges. With sufficiently high gain and the right output power from the intentional radiator (the AP and all cables, connectors, and components leading up to but not including the antenna), bridge links that span many miles or kilometers can be created.

The first highly directional antenna mentioned, the Yagi antenna, may be used for bridge links as well and is often used for shorter distance outdoor bridge links. However, Yagi antennas are also frequently used to transmit RF signals down long hallways or aisles in warehousing deployments where single stream data rates are sufficient as they are used in single-input-single-output (SISO) implementations. For example, with barcode scanners, high throughput rates are not required. In most cases, data rates at 11 or 12 Mbps and higher are sufficient, and many legacy devices will still connect at older 1- and 2-Mbps data rates. The Yagi antenna can help to provide coverage where it is needed without requiring excessive AP deployments.

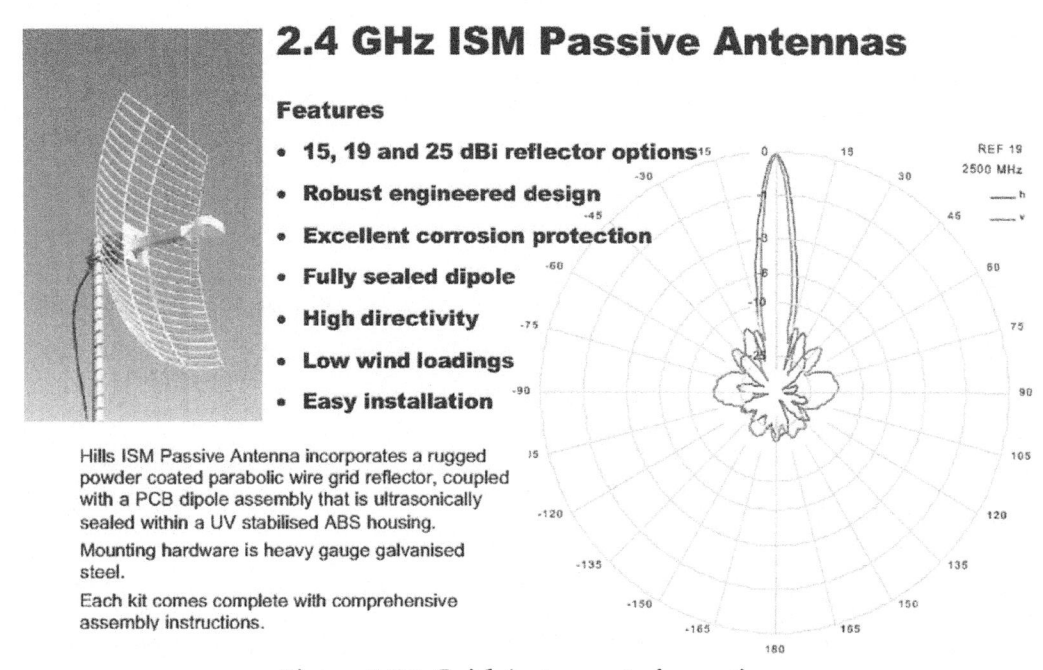

Figure 2.17: Grid Antenna Information

Figure 2.18 shows the antenna charts for a high-gain parabolic dish antenna. Notice the minimal rear lobes (the energy radiated in the unintended direction, also called side lobes). The vast majority of the RF energy is radiated in a narrow beam. The result is that the RF energy can travel great distances with sufficient signal strength for the reception.

The world record for a long-distance link using Wi-Fi technology was 310 kilometers in 2003. DEFCON has held competitions with many links created in excess of 30 miles. The point is that very long distance links can be created and maintained (though rarely in excess of a few miles or kilometers) using 802.11 equipment and appropriate antennas.

Vertical   **Horizontal**

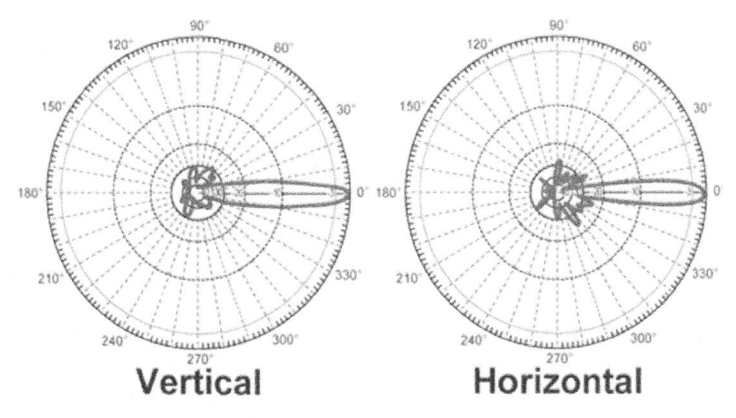

Figure 2.18: Parabolic Dish Antenna Chart

Internal vs. External Antennas

Our final discussion of antennas must address the issue of internal antennas (those within APs) and external antennas (those that attach to APs through cables and/or connectors). The most significant advantage of internal antennas is the reduced cost of implementation due to the lack of requirements to purchase external antennas and, in some instances, cables that need to be run to the antennas. The most significant advantage of external antennas is the ability to have better control over the distribution of RF signals in the coverage area.

In most indoor office environments, APs with internal antennas or APs that ship with external directly attached antennas are used. However, in high-density deployments that require support for hundreds or thousands of devices in a limited space, external directional antennas are often used. The choice between an internal or external antenna comes down to a balance of desired coverage and budgetary constraints.

> Remember that the primary reason for using external antennas indoors is to achieve a different coverage pattern than provided by internal AP antennas. With this exception, most indoor implementations use APs with internal or directly attached external antennas.

It is also useful to know that the vast majority of internal AP antennas have a gain of less than 4 dBi. External antennas will usually be required to achieve higher levels of gain. In fact, many APs have internal antenna gain of less than 3 dBi.

Additional RF Hardware

Finally, while not addressed in the CWS exam, several RF components may be used in a WLAN deployment. These include RF cables, connectors, amplifiers, and attenuators.

- RF cables are used to connect the AP or bridge to an antenna. They should be 50-Ohm cables, and the loss or attenuation rating of the cable, which is typically stated in loss per foot or meter, must be considered. An engineer needs this information to calculate EIRP.

- RF connectors are used to connect antennas to cables, cables to amplifiers and attenuators and antennas or cables to APs. They should be 50-Ohm connectors and match the connector type required by the device.

- Amplifiers are used to add active gain to the RF signal in a chain from the AP or bridge to the antenna. They should be rated for the appropriate frequency based on the signal, such as 2.4 GHz or 5 GHz. These devices should also be considered by the engineer in EIRP calculations.

- Attenuators are used to cause intentional attenuation, usually to comply with regulatory domain requirements.

Chapter Summary

In this chapter, you learned about the antennas used in RF-based WLANs. Different types of antennas were explored as well as the coverage patterns they provide. Finally, you briefly learned about the additional RF components used in WLANs. Now that you have learned the basics of RF and antennas from Chapters 1 and 2, the next chapter will now explore the 802.11 standard.

Points to Remember

Remember the following important points:

- Antennas radiate electromagnetic waves in the RF range in WLANs.

- Omnidirectional antennas uniformly radiate energy around the antenna element in a circular pattern.

- Omnidirectional antennas are usually placed in the center of the target coverage area.

- Semi-directional antennas radiate energy in a typical range between 50 and 180 degrees.

- Semi-directional antennas are usually placed at the edge of the target coverage area.

- Highly directional antennas radiate energy in very narrow propagation patterns and are useful for long-distance links.

- Highly directional antennas are usually placed at both ends of a bi-directional link.

- The azimuth (horizontal) chart shows the top-down view of an antenna's radiation pattern and indicates outward propagation.

- The elevation (vertical) chart shows the side view of an antenna's radiation pattern and indicates "up-and-down" propagation.

- Internal AP antennas are most commonly used for indoor office deployments.

- External antennas are used to provide specific coverage patterns in deployments and for bridge links and other special cases.

Review Questions

1. What two fields make up an RF wave?
   a. Radio and frequency
   b. Electric and magnetic
   c. Amplitude and phase
   d. Radio and Fresnel

2. Where are omnidirectional antennas typically placed within a target coverage area?
   a. Near the edge
   b. At the edge
   c. Central
   d. Outside coverage area

3. What kind of antenna is usually placed at the edge of the target coverage area?
   a. Parabolic dish
   b. Patch
   c. Dipole
   d. Omnidirectional

4. What unit is used to measure antenna gain?
   a. mW
   b. W
   c. dBi
   d. dBm

5. Which chart shows the top-down view of an antenna's propagation pattern?
   a. Azimuth
   b. Elevation
   c. E-Plane
   d. Vertical

6. When are external antennas used for indoor deployments?
   a. When a bridge link is being created spanning more than 3 kilometers
   b. When the coverage area exists around the AP on all sides
   c. When the antenna must be placed at the center of the target coverage area
   d. When specific coverage is needed

7. What type of antenna is typically internal to APs?
   a. Patch
   b. Panel
   c. Omnidirectional
   d. High gain

8. Which one of the following antennas would be more likely used in a 4-kilometer bridge link?
    a. Omnidirectional
    b. Patch
    c. Panel
    d. Grid

9. What kind of gain is created by antennas?
    a. Passive
    b. Active
    c. Attenuated
    d. Live

10. dBi compares a real antenna's gain to what theoretical antenna?
    a. Internal
    b. Iconic
    c. Intentional
    d. Isotropic

Review Answers

1. **B is correct.** The electric field forms the E-plane, and the magnetic field forms the H-plane.

2. **C is correct.** Omnidirectional antennas are usually placed near the center of the coverage area since they radiate out in all directions around the antenna.

3. **B is correct.** Patch and panel antennas are considered semi-directional and are often placed at the edge of a coverage area, for example, on a wall.

4. **C is correct.** dBi is preferred by engineers for the measurement of antenna gain. It is a relative value since the actual power depends on the amplitude of the input signal power and the metric is defined based on a comparison with an ideal radiator.

5. **A is correct.** The azimuth or horizontal chart shows the top-down view of the radiation pattern.

6. **D is correct.** Most internal office deployments used internal AP antennas. When specific coverage patterns are required, external antennas may be used.

7. **C is correct.** Most APs use omnidirectional antennas internally.

8. **D is correct.** Bridge links over greater distances usually use a parabolic dish or grid antenna.

9. **A is correct.** Antennas create passive gain because they do not increase the amplitude of the RF signal received, but instead they focus it.

10. **D is correct.** An isotropic radiator is an ideal theoretical antenna that radiates energy equally in all directions in a spherical pattern.

Chapter 3 – The 802.11 Standard

Objectives Covered

2.1 Know the frequency bands used by 802.11
    2.1.1   2.4 GHz – 802.11b/g/n
    2.1.2   5 GHz – 802.11a/n/ac
    2.1.3   Sub-1 GHz – 802.11ah
    2.1.4   60 GHz – 802.11ad

2.2 Identify Physical Layer (PHY) characteristics
    2.2.1   Data rates
    2.2.2   Bands used
    2.2.3   Supported technologies (laptops, tablets, video devices, Internet of Things (IoT))

In computer networking, standards define the communication procedures that are used by compatible devices. Standards exist at the hardware level (e.g., Ethernet, Wi-Fi) and at higher levels (e.g., TCP-IP, HTTP, FTP). The hardware-level standards used by modern networking equipment are created by the Institute of Electrical and Electronics Engineers (IEEE). The higher-level standards (or protocols) are created by the Internet Engineering Task Force (IETF) through Request for Comments (RFC) documents. This chapter is focused on the IEEE 802.11 standard and highlights two other IEEE standards that are important for WLAN deployments.

IEEE Standards

IEEE standards are created through a drafting process that results in collaborative standards for network communications. This means that standards go through a development process consisting of multiple drafts before the final standard is ratified. Before looking at the 802.11 standard itself, the development process for standards will be explored. Then, the 802.11 standard will be covered in sufficient detail to create a foundation on which you can build your knowledge. Finally, the Ethernet standard (802.3) and port security standard (802.1X) will be briefly reviewed, since they are both highly relevant to WLANs.

Standards Development

IEEE standards are managed by working groups. For example, there is an 802.3 working group and an 802.11 working group. The working group oversees the creation and maintenance of the standard. When the initial standard is created, several drafts are generated, and feedback is received and incorporated as needed in the drafting process. When the final draft is ratified (approved by a vote of active members), it becomes a standard.

After a standard exists (i.e., it has been ratified), it must be maintained. Figure 3.1 illustrates the lifecycle for standards used by the IEEE. Draft amendments are created and may go through several drafts. When a draft amendment is ratified, it becomes part of the standard. A ratified amendment may add features to the standard or it may add entirely new ways of communicating on the network (known as physical layers [PHYs]). For example, 802.11n was an amendment that added the High Throughput (HT) PHY to the standard.

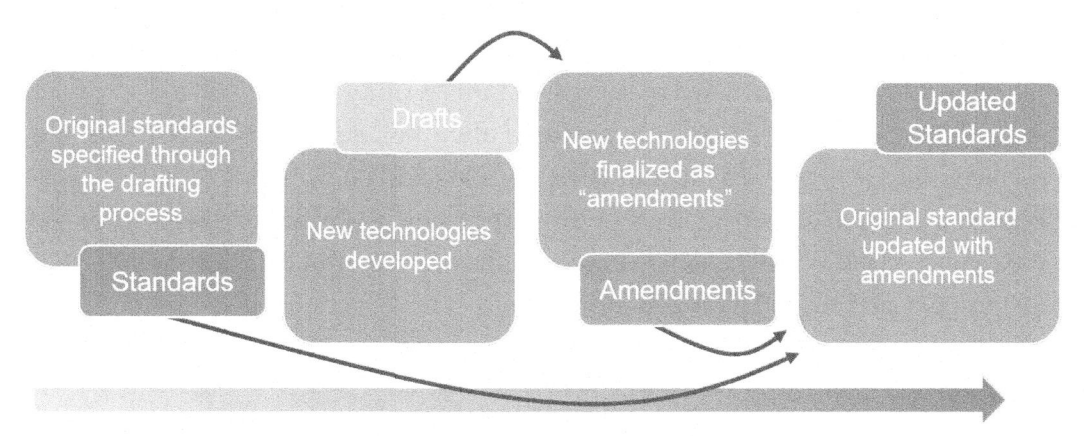

Figure 3.1: The Standards Lifecycle

The phrase "802.11 as amended" refers to the most recent revision document (currently 802.11-2016) plus any ratified amendments released after the revision. Therefore, 802.11ah modified 802.11-2016 to add the Sub-1 GHz (S1G) PHY to the standard, and the standard is actually 802.11-2016 plus the changes introduced in the 802.11ah amendment.

Figure 3.2 shows the 802.11-2016 standard, which notes that it is a revision of 802.11-2012. In fact, the 802.11-2016 standard includes all revisions before 802.11-2012 plus the following amendments:

- 802.11ae-2012 – Prioritization of Management Frames
- 802.11aa-2012 – MAC Enhancements for Robust Audio Video Streaming
- 802.11ad-2012 – Enhancement for Very High Throughput in the 60 GHz Band
- 802.11ac-2013 – Enhancements for Very High Throughput for Operation in Bands below 6 GHz
- 802.11af-2013 – Television White Spaces (TVWS) Operation

Because the 802.11ah amendment was not ratified in time to be included in the 802.11-2016 revision, it will apply as an amendment, and the engineer must compare the 802.11ah amendment to the 802.11-2016 revision to fully understand the changes made.

802.11

With an understanding of the IEEE standards development process, you can better understand all the terminology you see related to 802.11 wireless networks. The latest revision of the standard at the time of this writing is 802.11-2016. Amendments that have been either ratified after the release of 802.11-2016 or are forthcoming include 802.11ah and 802.11ax. One adds new bands for 802.11 operations (802.11ah), and the other provides a new PHY for 802.11 devices (802.11ax). The CWS-100 exam is based on the features and functions available in 802.11-2016 and the 802.11ah amendment.

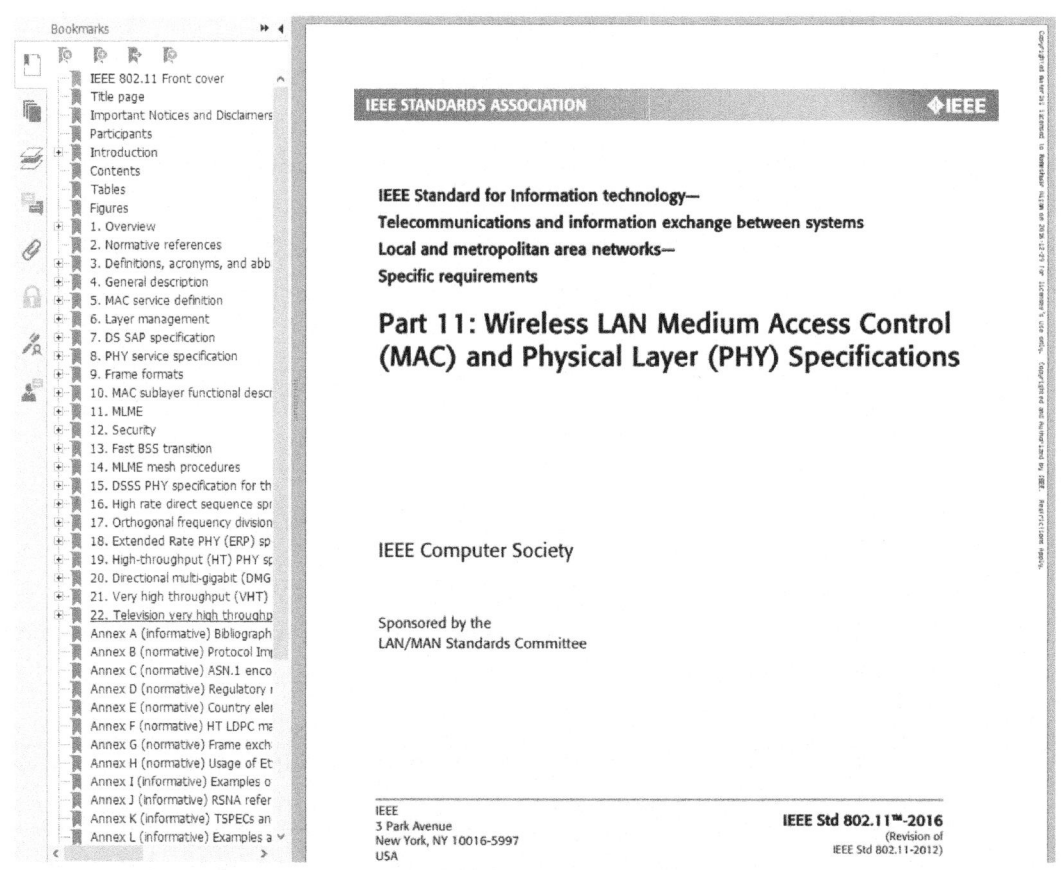

Figure 3.2: 802.11-2016 Standard

After this section, the remainder of this chapter will explore the 802.11 standard in more detail. Beginning with the section "Frequency Bands Used by 802.11 Devices", the 802.11 standard will be explained in greater detail. For now, it is important to explore other IEEE standards.

802.3

The 802.3 standard defines wired Ethernet networking. It is the most commonly used networking standard for desktop computers, connected laptops, servers and network infrastructure devices (e.g., routers, switches). Nearly all 802.11 wireless APs connect to Ethernet networks and bridge wireless clients onto the network. Modern APs have one or two gigabit Ethernet ports for connectivity, and most enterprise APs also take advantage of the Power over Ethernet (PoE) specification defined in the 802.3 standard. Therefore, both the networking specifications and the power provisioning specifications in 802.3 apply to wireless networks today.

802.1X

The 802.1X standard defines port-based network access control, which is sometimes called port-based authentication, though some vendors use the phrase port security as a vendor-specific term. 802.1X defines two virtual ports that exist in each physical port. One virtual port is called the controlled port, and the other is the uncontrolled port. The controlled port cannot be used until authentication has been successfully completed on the uncontrolled port. The uncontrolled port allows only authentication-related traffic, and the controlled port once enabled allows all other traffic. 802.11 WLANs can use 802.1X for secure access control when using WPA- or WPA2-Enterprise security.

> Even though 802.11 devices do not connect to an AP with a wire, they still have a connection, and that connection in the AP is considered a port. Therefore, just as a physical Ethernet port can apply 802.1X, an AP can apply 802.1X to an association with a client.

Figure 3.3 shows the basic flow of 802.1X authentication for a wireless client device. The client is called the supplicant because it must make supplication to an authentication server to gain access. The AP or WLAN controller is called the authenticator in 802.1X terminology, and it acts as the proxy or go-between for the supplication and the authentication server. The final piece is the authentication server, which is typically a RADIUS (Remote Authentication Dial-In User Service) server. RADIUS is an IETF standard itself. In addition to this, some Extensible Authentication Protocol (EAP) method is used for authentication. EAP methods will be discussed more in Chapter 5 (*WLAN Security, BYOD and Guest Access*).

As a CWS certified professional, you should know the basic components required to implement a secure WLAN. These components include the supplicant, authenticator, and authentication server. You should also know that the authentication server is most often a RADIUS server. The RADIUS server may use an internal database, or it may use an external database for user authentication. For example, many RADIUS servers will connect to Microsoft Active Directory Domain Services (AD DS) to validate user credentials during the authentication process.

> *EAP* is an authentication framework defined in an IETF RFC document (RFC 3748). A specific implementation of EAP is called an EAP method or EAP type. It provides enterprise authentication to modern WLANs in concert with 802.1X.

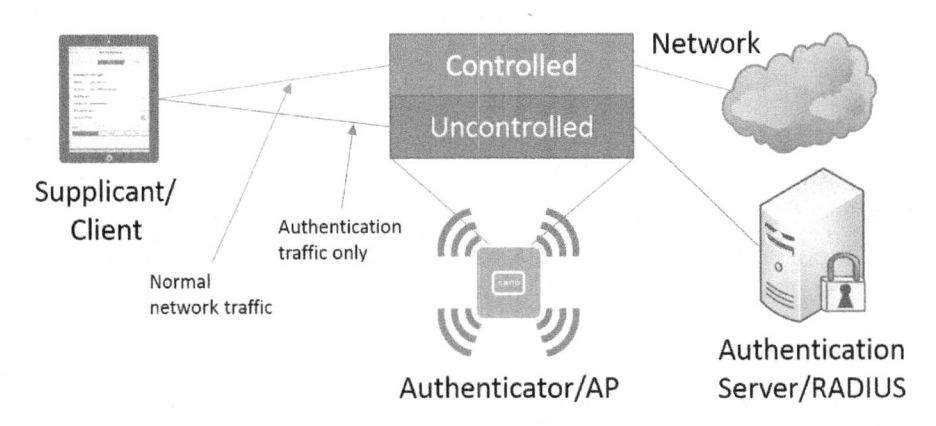

Figure 3.3: 802.1X Port-Based Authentication

Frequency Bands Used by 802.11 Devices

802.11 wireless devices can operate in one of four primary frequency bands:

- Sub-1 GHz (S1G)
- 2.4 GHz
- 5 GHz
- 60 GHz

Traditional WLAN devices operate in either the 2.4 GHz or 5 GHz frequency bands. These devices include laptops, tablets, mobile phones, wireless APs and many other specialty devices (e.g., video cameras, door locks, mobile VoIP handsets, push-to-talk devices, wireless video systems). 802.11b, 802.11g, and 802.11n devices can all operate using the 2.4 GHz frequency band and are backward compatible with each other. 802.11a, 802.11n, and 802.11ac devices can all operate using the 5 GHz frequency band and are compatible with each other. 2.4 GHz-only devices cannot communicate with 5 GHz-only devices. Therefore, a device operating as a 2.4 GHz-only 802.11n device cannot communicate with a device operating as a 5 GHz 802.11n device, because they communicate with different frequencies.

The frequency ranges used for 2.4 GHz and 5 GHz are as follows:

- 2.4 GHz uses the range from 2400 MHz (2.4 GHz) to 2500 MHz (2.5 GHz), and the actual usage range for 802.11 channels is from 2.401 GHz to 2.495 GHz

- 5 GHz uses the range from 5000 MHz (5 GHz) to 5835 MHz (5.835 GHz), and the actual usage range for 802.11 channels is from 5.170 GHz to 5.835 GHz, though not all areas within this range are used.

Within these ranges, specific portions are defined as channels of 22 MHz for the oldest wireless devices and 20 MHz or some factor of 20 MHz (e.g., 40 MHz, 80 MHz, and 160 MHz) for newer devices. Channels are discussed in more detail in Chapter 4 (*Channel Plans and Performance Factors*).

The 2.4 GHz and 5 GHz bands are used differently depending on a country's radio authorities and their adoption of a set of rules called the regulatory domain. It is important to know which frequencies are available in the regulatory domain where the WLAN will be installed. As a CWS, you may want to pay attention to this to avoid ordering mishaps that could prevent your customer from taking full advantage of the products you are offering.

> The 2.4 GHz and 5 GHz bands are used differently in varying regulatory domains. It is important to know what frequencies are available in the regulatory domain of the installed WLAN. This regulatory information is available online but is not part of the CWS exam.

The S1G bands are used for both 802.11ah devices (also called Wi-Fi HaLOW) and 802.11af devices (TV white spaces). The specific frequency ranges used vary based on regulatory domains for 802.11ah and available frequency space for 802.11af. More details about these bands will be provided in the next section of this chapter (*Physical Layers in 802.11*).

The 60-GHz band, with respect to the 802.11 standard, is used only for the Directional Multi-Gigabit (DMG) PHY and is not covered in detail in this book or the CWS exam. However, it is mostly used with video devices and other short-range wireless devices that need very high throughput rates but require short-distance links. The DMG PHY was first defined in 802.11ad and is now part of the 802.11-2016 standard. It is basically an implementation of 802.11ac (VHT) in the 60-GHz band.

Physical Layers in 802.11

The 802.11-2016 standard defines several different physical layers (PHYs) that provide different data rates, channel widths, and operational frequency bands. Additionally, the 802.11ah draft amendment defines an added PHY, and more PHYs will be added in the future. The CWS exam requires that you are aware of the basic features of the PHYs defined in 802.11-2016 and 802.11ah, including the following:

- Data rates (the speed of Wi-Fi frame transmissions supported by the PHY)
- Bands used (the frequency bands supported by the PHY)

- Supported technologies (for example, the devices that may use a particular PHY)

Remember that data rates and throughput are two very different things in wireless networks. The data rate is the rate at which data is transmitted on the wireless medium (RF channel). Throughput is the rate at which higher layer data (usually TCP or UDP) is transmitted and is often 20-60% less than the wireless link data rate.

The channel width and modulation used significantly impact the actual data rates that are available. Each PHY supports specific data rates based on the combination of channel width, modulation, coding and a few other features. The data rates are not in some way arbitrarily variable (for example, going from 11 Mbps to 10.9 Mbps to 10.8 Mbps, and so forth), but they are specific data rates supported based on combinations of these factors (for example, going from 11 Mbps to 5.5 Mbps to 2 Mbps, and so forth).

The following subsections provide a brief overview of each PHY with all of the information that a CWS should understand. This information is sufficient for the exam and can be used to explain the technologies to decision makers and to make purchasing decisions.

> A *physical layer (PHY)* is a specification for communications of data bits across the RF medium. It defines how modulation and coding work in communications and is focused on moving ones and zeroes between two devices.

DSSS

The oldest PHY still supported by modern 802.11 devices is the *Direct Sequence Spread Spectrum (DSSS)*. DSSS uses a 22-MHz wide channel and operates only in the 2.4 GHz band. Each channel is assigned based on a channel center frequency (e.g., 2.412 GHz for channel 1) and uses 11 MHz on either side of the center frequency. Therefore, channel 1 would use the range from 2.401 to 2.423 for the 22 MHz channel.

As with all PHYs introduced before 802.11n (HT), DSSS only supports one spatial stream. This means that the transceiver (transmitter/receiver) sends one stream of data and receives one stream of data at a time. More recent PHYs will support multiple streams for transmission and reception, significantly increasing the available data rates.

Remembering the supported data rates for DSSS is simple. Only two data rates are supported: 1 Mbps or 2 Mbps. By today's standards, this PHY is very slow. However, the DSSS PHY is supported by all 802.11 devices that operate in the 2.4 GHz band, including the newest 802.11n devices.

In summary, the DSSS PHY supports data rates of 1 or 2 Mbps. It operates in the 2.4 GHz band and only supports a single spatial stream. All DSSS transmissions use a 22-MHz channel width. Figure 3.4 provides a summary of this information.

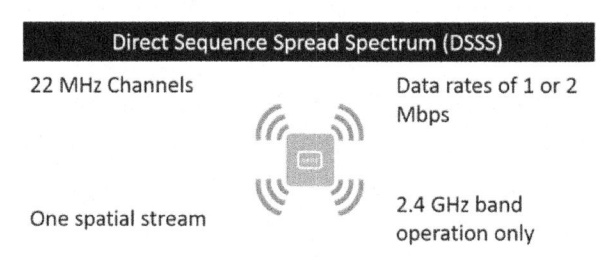

Figure 3.4: DSSS PHY Summary

HR/DSSS

The *High Rate/Direct Sequence Spread Spectrum (HR/DSSS)* PHY was released with the 802.11b amendment in 1999. It introduced more advanced modulation techniques, allowing for data rates of 5.5 and 11 Mbps while still supporting the DSSS data rates of 1 and 2 Mbps. HR/DSSS uses the same 22-MHz wide channels as DSSS and only supports a single spatial stream. Like DSSS, HR/DSSS only operates in the 2.4 GHz frequency band.

All newer PHYs operating in the 2.4 GHz frequency band are designed to be backward compatible with earlier PHYs in the same band. This statement means that an HR/DSSS device can communicate with a device that only supports DSSS. Additionally, though the details of the HT PHY have not yet been explored, an HT

PHY device can communicate with a DSSS PHY device as long as they both operate in the same 2.4 GHz frequency band.

> WLAN administrators may disable backward compatibility by disallowing lower data rates. However, this is a configuration constraint and not a radio or device constraint. With all data rates enabled, newer 2.4 GHz 802.11 devices can communicate with all older devices.

Figure 3.5 provides a summary of information related to the HR/DSSS PHY.

High Rate Direct Sequence Spread Spectrum (HR/DSSS)

22 MHz Channels

Data rates of 1, 2, 5.5 or 11 Mbps

One spatial stream

2.4 GHz band operation only

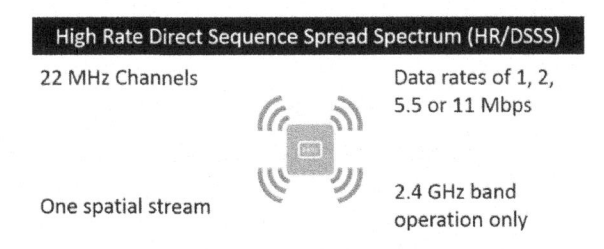

Figure 3.5: HR/DSSS PHY Summary

OFDM

The *Orthogonal Frequency Division Multiplexing (OFDM)* PHY was the first to support 5 GHz band operations. This PHY was made available through the 802.11a amendment in 1999. In addition to 5 GHz band support, OFDM was the first to use 20-MHz channels instead of 22-MHz channels. All modern PHYs that are based on OFDM use some factor of 20-MHz channels. For example, they may use 20-MHz (802.11a/n/ac) or 40-MHz (802.11n/ac) channels (as well as 80 and 160 MHz in 802.11ac). 802.11a was the first PHY amendment to use OFDM modulation, and the PHY is named after the modulation. All PHYs introduced since 802.11a also use OFDM modulation, but they have a different PHY name to differentiate them from 802.11a OFDM clearly.

OFDM still uses one spatial stream, but with enhanced modulation, it supports data rates of 6, 9, 12, 18, 24, 36, 48 and 54 Mbps. Notice that it does not support 1, 2, 5.5 or 11 Mbps. OFDM operates in the 5 GHz frequency band and has no need to be backward compatible with DSSS or HR/DSSS.

> The 802.11ax amendment will introduce a new modulation called OFDMA, but it has not been ratified as of writing this chapter. The PHY name currently proposed for 802.11ax is High Efficiency (HE), but this may change by the time of ratification.

Orthogonal Frequency Division Multiplexing (OFDM)

20 MHz Channels

Data rates of 6, 9, 12, 18, 24, 36, 48 or 54 Mbps

One spatial stream

5 GHz band operation only

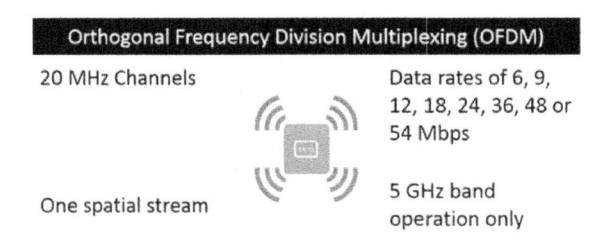

Figure 3.6: OFDM PHY Summary

ERP

The *Extended Rate PHY (ERP)* was introduced to extend OFDM modulation down into the 2.4 GHz band. The implemented PHY features in 802.11 devices use the same OFDM modulation used in 5 GHz 802.11a devices and use 20-MHz channels. There are some slight differences in the way that the PHY was implemented, but for the CWS, it is sufficient to know that it provides the same basic functionality as OFDM provided in the 5 GHz frequency band.

> Remember that ERP is the first OFDM-based PHY that operates at 2.4 GHz, and OFDM is the first OFDM-based PHY that operates at 5 GHz. HT and VHT, which will be explained later in this section, are both OFDM-based as well.

Operating at 2.4 GHz, all ERP devices (which are also called 802.11g devices based on the amendment that defined ERP) support backward compatibility with HR/DSSS and DSSS PHY devices. This fact is another characteristic that differentiates it from OFDM PHY at 5 GHz. OFDM was the first PHY at 5 GHz and required no backward compatibility. To accomplish backward compatibility, ERP (802.11g) devices still support the DSSS data rates of 1 and 2 Mbps and the HR/DSSS data rates of 5.5 and 11 Mbps. In addition, they support the same data rates as OFDM (802.11a) of 6, 9, 12, 18, 24, 36, 48 and 54 Mbps. To be clear, the ERP PHY supports only the data rates of 6, 9, 12, 18, 24, 36, 48 and 54 Mbps, but all devices implementing the ERP PHY also implement the DSSS and HR/DSSS PHYs so that the 1, 2, 5.5 and 11 Mbps data rates are also supported.

Figure 3.7 provides a summary of information for the ERP PHY and the essential characteristics to know as a CWS professional. Note that the data rates listed are those added by the ERP PHY, but remember that ERP or 802.11g devices support backward compatibility by also implementing the DSSS and HR/DSSS PHYs and the data rates they support.

Extended Rate PHY (ERP)

20 MHz Channels

Data rates of 6, 9, 12, 18, 24, 36, 48 or 54 Mbps

One spatial stream

2.4 GHz band operation only

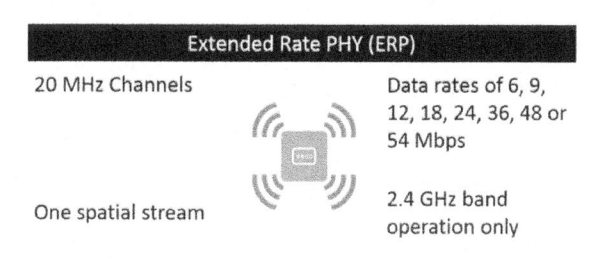

Figure 3.7: ERP PHY Summary

HT

The *High Throughput (HT)* PHY was introduced in the 802.11n amendment and offers several advantages over older PHYs. HT provides wider channels by combining two 20-MHz sections into a 40-MHz channel. Therefore, HT provides either 20-MHz or 40-MHz channels. An AP that offers a 40-MHz channel can still service 20-MHz clients on its primary channel.

The primary channel will be one of the defined channel numbers, such as 1, 6, 11, 36 or 44. Then, the secondary channel, which provides a total of 40 MHz, will be 20 MHz above or below the primary channel. When a device connects to an AP that offers a 40-MHz channel, and the connecting device supports only a 20-MHz channel, it will communicate with the AP using the primary channel. The 40-MHz client devices can use the entire 40-MHz channel.

Wider channels result in higher data rates even with no additional features. However, the HT PHY also introduces the capability to use multiple spatial streams through *Multiple Input/Multiple Output (MIMO)*. MIMO takes advantage of RF propagation behaviors to send multiple concurrent streams of data from the transmitter to the receiver. The HT PHY supports up to four spatial streams; however, most devices support from one to three spatial streams when using the HT PHY.

> 40 MHz channels should never be used at 2.4 GHz even though most APs allow for it. As you will learn in the next chapter, the 2.4 GHz band lacks the frequency space to allow for 40-MHz channels in enterprise deployments.

Another unique feature of the HT PHY is that it operates at either 2.4 GHz or 5 GHz. More channels are available in the 5 GHz band, so it is the preferred band, but many devices operate only at 2.4 GHz (even some of the newest devices being sold), so this band continues to be supported in nearly all implementations.

Finally, the HT PHY offers many more data rate possibilities than earlier PHYs. The available data rates will depend on the channel width (20 MHz vs. 40 MHz), the number of spatial streams and the modulation and coding used. Some additional factors impact the available data rates but are beyond the scope of the CWS exam or the knowledge required to perform the job role. The maximum data rate achievable with the HT PHY, assuming a 40-MHz channel and the highest modulation and coding rate, is 600 Mbps. Most HT or 802.11n devices support maximum data rates of 150, 300 or 450 Mbps because the devices support from one to three spatial streams, but the standard allows for up to 600 Mbps.

It is beneficial to know that 2.4 GHz devices will support maximum data rates of 72.2, 144.4 and 216.7 Mbps (sometimes these numbers are rounded to 72, 144 and 217 Mbps) because they will only use 20-MHz channel widths in a proper implementation. While 2.4 GHz devices could be configured to support the higher data rates offered by 40-MHz channels, they should not be. When using 40-MHz, 2.4 GHz channels in a multi-AP deployment, the degradation in performance due to channel overlap is not worth the gains offered by 40-MHz channels. However, in 5 GHz standard deployments, 40-MHz channels can be beneficial, depending on the type of network being deployed. The next chapter will provide more details on this issue. Figure 3.8 shows a summary of information for the HT PHY. The information in the figure is the primary information you should remember for the CWS exam.

High Throughput (HT)

20 or 40 MHz Channels

Data rates up to 600 Mbps

Up to four spatial streams

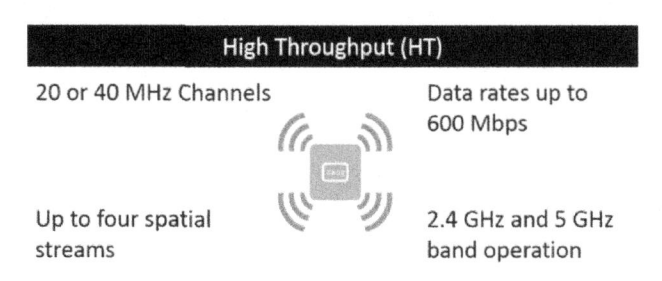

2.4 GHz and 5 GHz band operation

Figure 3.8: HT PHY Summary

VHT

The *Very High Throughput (VHT)* PHY moves 802.11 networks even further than the HT PHY. The VHT PHY now supports additional channel widths of 80 MHz and 160 MHz (though 160-MHz channels should not be used in enterprise deployments, and 80-MHz channels should rarely be used). The base channel width is still 20 MHz, but two, four or eight 20-MHz portions may be used to form the wider channels.

The wider channels achieve higher data rates, but VHT (802.11ac) also adds support for more spatial streams. A VHT device can use up to eight spatial streams. The first devices that were released supported three spatial streams, but devices that support four spatial streams are now on the market. Whether eight spatial streams will be used is yet to be seen, because the general trend in client devices is to stay with fewer spatial streams, which reduces battery consumption. It is imperative to know that the VHT PHY works only in the 5 GHz frequency band. There is no support for VHT in the 2.4 GHz band, unlike the HT PHY. The primary reason for this decision to limit VHT to the 5 GHz band was the lack of frequency space for wider channels at 2.4 GHz.

Finally, VHT devices can achieve a maximum data rate of 6933.3 Mbps; however, this data rate would require eight spatial streams. Because 802.11ac devices implement no more than four spatial streams today, the real-world peak data rate is 3466.7 Mbps. To achieve this data rate of 3466.7 Mbps, the AP and client must both support four spatial streams and use a 160-MHz channel. Given the reality that few 802.11ac APs will be implemented with channels configured with a bandwidth of more than 40 MHz, it is more likely that you will see maximum data rates of 800 Mbps for four spatial streams on a 40-MHz channel.

Always remember that the data rate available for a link is constrained by the least capable component of that link. For example, if an AP is configured with a 40-MHz channel and supports four spatial streams, a four spatial stream client supporting a 40-MHz channel could potentially connect with a data rate of 3466.7 Mbps. However, a single stream, 40-MHz client will connect to the same AP with a maximum data rate of 200 Mbps. As you can see, the real world is often very different from marketing literature. Figure 3.9 provides a summary of this key information related to the VHT PHY.

Very High Throughput (VHT)

20, 40, 80 or 160 MHz Channels

Up to eight spatial streams

Data rates up to 6933.3 Mbps

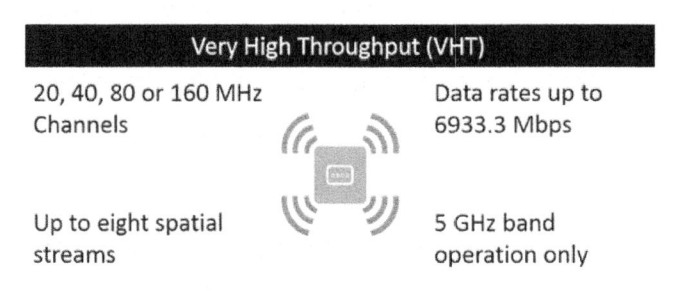

5 GHz band operation only

Figure 3.9: VHT PHY Summary

TVHT

The *Television High Throughput (TVHT)* PHY is not included on the CWS exam beyond a general awareness of its target use and frequencies used, but information related to its capabilities is provided here for completeness. This PHY is designed to take advanced of unused frequencies in the bands often used for television and other broadcasts. Because it is designed to use such spaces, it supports very narrow channel widths of 6, 7 or 8 MHz, depending on the regulatory domain in which it operates. Additionally, the channel widths of 6, 7 or 8 MHz (called Basic Channel

Units (BCUs)) can operate as 1, 2 or 4 BCUs. Therefore, with two 7-MHz BCUs, the total frequency space available for the transmissions would be 14 MHz.

The maximum data rate supported with the use of four 8-MHz BCUs (32 MHz of frequency space) is 568.9 Mbps. This is accomplished with four spatial streams. Like 802.11n devices, TVHT devices can use one to four spatial streams. Finally, TVHT operates in the frequency range from 50 MHz to 790 MHz and uses frequency space as allocated by the operating regulatory domain. Figure 3.10 provides a summary of this information related to TVHT.

Television Very High Throughput (TVHT)

Basic Channel Unit (BCU) of 6, 7 or 8 MHz with combinations of 1, 2 or 4 BCUs

Up to four spatial streams

Data rates up to 568.9 Mbps

TV whitespace usage from 50 MHz to 790 MHz

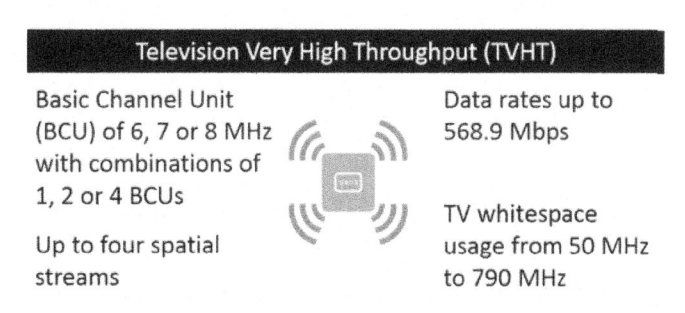

Figure 3.10: TVHT PHY Summary

S1G

The *Sub-1 GHz (S1G)* PHY was designed with long-range, low-data-rate communications in mind and is defined in the 802.11ah amendment. It is ideal for Internet of Things (IoT), industrial automation and monitoring networks. The S1G PHY operates on 1-, 2-, 4-, 8- or 16-MHz channels, and it appears likely that more devices will use the 1-, 2- and 4-MHz channels since their likely use cases do not warrant higher data rates.

The maximum data rate supported on the S1G PHY is 346.6667 Mbps. This rate is based on a 16-MHz channel and four spatial streams. Given the desire for extended battery life and the low need for high data rates, (e.g., devices supporting the S1G PHY), we are likely to continue to see many single-stream devices. Such devices, operating on a likely maximum of 4- or 8-MHz channels, will achieve a maximum data rate of 8666.7 Kbps for a 2-MHz, single-stream device or 20,000 Kbps (20

Mbps) for a 4-MHz, single-stream device. The actual frequencies used will vary by regulatory domain but will all be less than 1 GHz. Some will operate in the 700-MHz range, others in the 900-MHz range and still others in between. For the CWS, it is important to remember that all S1G PHY devices operate at frequencies below 1 GHz. Figure 3.11 provides a summary of this key information related to the S1G PHY.

Sub-1 GHz (S1G)

1, 2, 4, 8 and 16 MHz channel widths

Data rates up to 346.6667 Mbps

Up to four spatial streams

Frequencies below 1 GHz varying by regulatory domain

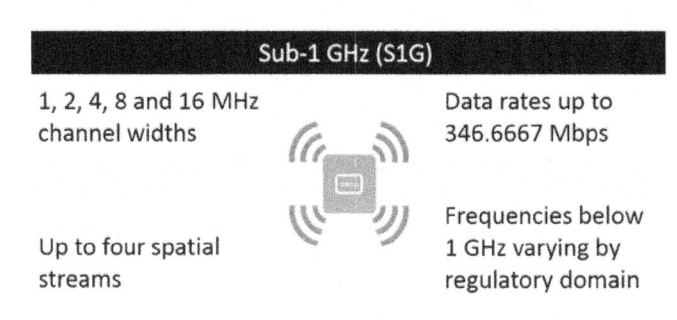

Figure 3.11: S1G PHY Summary

Table 3.1 provides a summary of the essential details related to the various 802.11 PHYs. Remember that the TVHT PHY information will not be tested on the CWS exam but is provided for reference.

Key 802.11 Standard Non-PHY Amendments

This final section on the 802.11 standard will review key amendments to the standard that were ratified in the years following the initial release in 1997 and that were not intended to introduce new PHYs. This includes security enhancements, operational changes, and additional features. The following list comprises the most significant amendments of which the CWS candidate should be aware:

- **802.11i-2004:** 802.11i introduced enhanced security to resolve weaknesses in the original security protocols implemented in 802.11-1997. The original security was Wired Equivalent Privacy (WEP), and it had many flaws that were resolved by 802.11i. 802.11i, which is the foundation for the Wi-Fi Alliance certifications known as WPA and WPA2.

PHY Layer	Max Data Rate	Max Spatial Streams	Band	Max Channel Width
DSSS	2 Mbps	1	2.4 GHz	22 MHz
HR/DSSS	11 Mbps	1	2.4 GHz	22 MHz
OFDM	54 Mbps	1	5 GHz	20 MHz
ERP	54 Mbps	1	2.4 GHz	20 MHz
HT	600 Mbps	4	2.4 GHz/5 GHz	40 MHz
VHT	6933.3 Mbps	8	5 GHz	160 MHz
TVHT	568.9 Mbps	4	54 to 790 MHz TV Whitespace only	32 MHz
S1G	346666.7 Kbps (346.6667 Mbps)	4	Sub-1 GHz – specifics vary by regulatory domain	16 MHz

Table 3.1: PHY Layers and Specifications

- **802.11e-2005:** 802.11e introduced probabilistic quality of service (QoS) to 802.11 networks. It provides markings that identify the priority of the WLAN traffic so that devices can forward higher priority traffic across the wireless medium before lower priority traffic. Several other performance-related enhancements were also introduced with this amendment that impact power management and multimedia communications.

- **802.11k-2008:** 802.11k provided radio resource measurement for WLANs. This enhancement allows the wireless network to gather information from its clients and neighbors and better understand the conditions on the network so that automated management decisions may be made.

- **802.11r-2008:** 802.11r introduced new protocols for secure roaming from one AP to another in the same larger wireless network. The newest APs today often support 802.11r, but the client market has not yet been saturated.

- **802.11w-2009:** 802.11w addressed a security weakness that was not addressed by 802.11i, which was the potential for denial of service (DoS) attacks through management frame manipulation. Management frame protection (MFP) was introduced by 802.11w, making it harder to deauthenticate or disassociate a valid station from the network.

- **802.11u-2011:** 802.11u provides enhancements to the 802.11 standard in the area of interworking with external networks. Enhanced network discovery and selection options are made available through the Access Network Query Protocol (ANQP). This amendment provided the primary enhancements utilized by the Wi-Fi Alliance Passpoint certification, which is also known as HotSpot 2.0.

> Be sure to remember these basic amendment descriptions for exam day. You may be asked questions related to the capabilities introduced by a given amendment, and the previous list covers the testable amendments.

802.11 Networking Terminology

The CWS should understand terms related to 802.11 wireless networking from the standard, including the following:

- Basic Service Set (BSS), including infrastructure BSS (called BSS for short) and independent BSS (called IBSS for short)
- Service Set Identifier (SSID)
- Basic Service Area (BSA)
- Extended Service Set (ESS)

A BSS is a set of interconnected 802.11 devices. The infrastructure BSS is the network of devices connecting with a single AP. The BSS is identified by a BSS Identifier (BSSID) under the hood, but it is better known to users by the SSID. The SSID is an alphanumeric name that makes connecting to the BSS easier because the BSSID is a hexadecimal value. All devices in a BSS communicate through the AP and, while they can communicate directly with other members of the same BSS in some situations, they do not usually communicate directly with each other. Figure 3.12 illustrates the concept of a BSS.

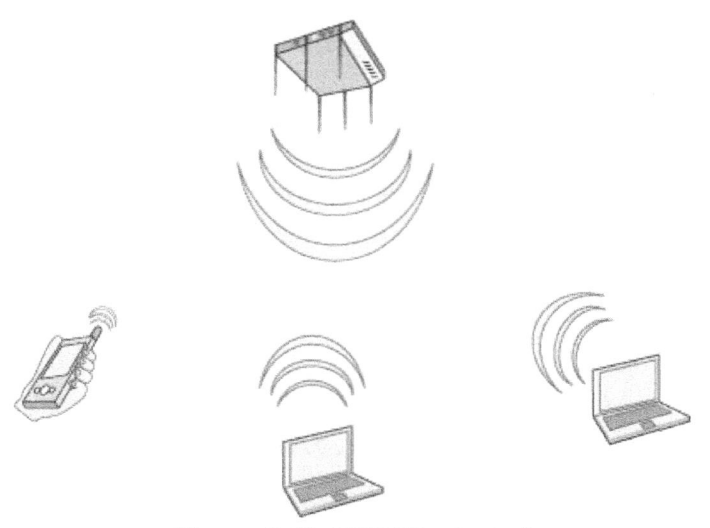

Figure 3.12: BSS Illustrated

An independent BSS (IBSS) is a set of interconnected devices that communicate directly with each other and in range of each other without the use of an AP. It is also called an ad-hoc network. Many enterprise environments banned the use of IBSSs due to the potential security concerns of inexperienced users implementing such networks. Figure 3.13 illustrates the concept of an IBSS.

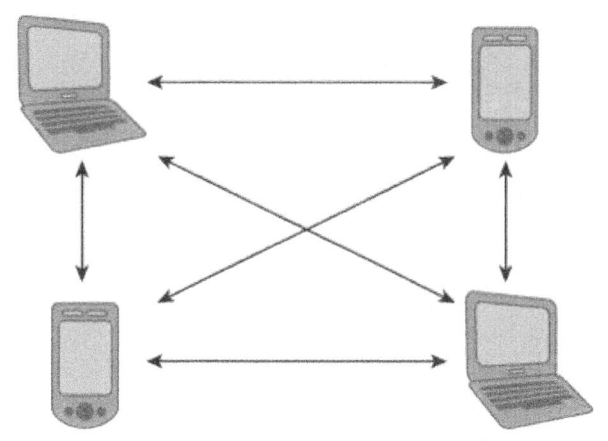

Figure 3.13: IBSS Illustrated

The BSA is the area covered by a BSS. Stated differently, it is the area within which client devices can connect to the AP. The term cell is often used in everyday communications among engineers instead of the 802.11 term BSA. Both terms refer to the same concept.

Finally, an ESS is technically defined in the standard as one or more interconnected BSSs. However, in practical deployment, an ESS is multiple BSSs that interconnect through a shared distribution system medium (DSM). To simplify the topic, a DSM is a shared network to which all BSS APs connect. For example, if three APs operate in the same ESS, they are likely connected to the same Ethernet network. In nearly all cases, all BSSs in an ESS use the same SSID. The SSID is not required to be the same because some other element could be used to identify coexistence in the ESS, but for all practical purposes, the SSID is always used. Figure 3.14 illustrates an ESS.

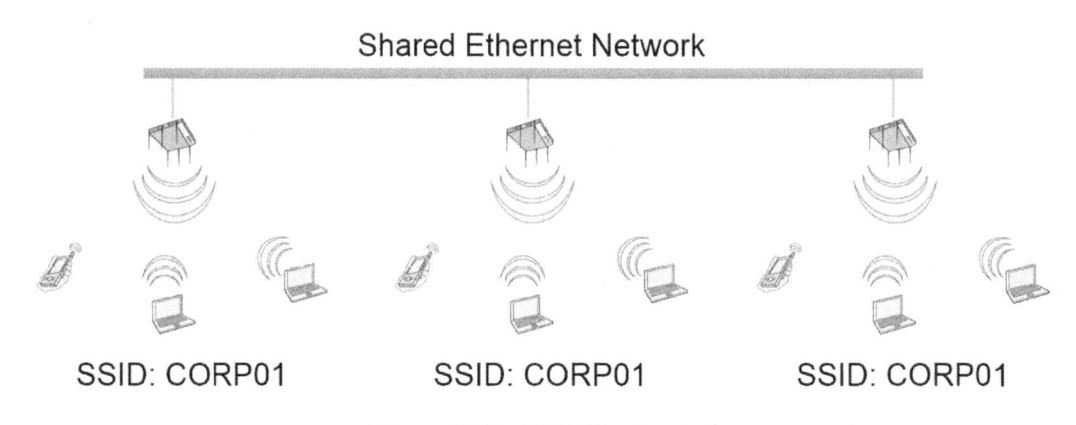

Figure 3.14: ESS Illustrated

Chapter Summary

In this chapter, you learned about IEEE standards in general and the 802.11 standard specifically. You reviewed each PHY supported by the standard as well as the data rates, frequency bands and additional features implemented or used by each PHY. Finally, you explored various amendments to the standard and the capabilities they introduced. In the next chapter, you will focus on channel planning and efficient implementation of WLAN devices.

Points to Remember

Remember the following important points:

- IEEE standards are developed by working groups and go through draft phases before they become a standard.

- The original 802.11 standard was released in 1997, and the only PHY remaining in production from that initial standard is DSSS.

- The DSSS PHY supports a maximum data rate of 2 Mbps; it also has one lower data rate (1 Mbps) and operates in the 2.4 GHz band.

- The HR/DSSS PHY introduced the additional data rates of 5.5 and 11 Mbps and operated in the 2.4 GHz band.

- Both DSSS and HR/DSSS use 22-MHz channels, but ERP and OFDM use 20-MHz channels.

- HT supports both 20- and 40-MHz channels and VHT supports 20-, 40-, 80- and 160-MHz channels.

- The OFDM PHY operates at 5 GHz and supports data rates up to 54 Mbps.

- The ERP PHY operates at 2.4 GHz and supports data rates up to 54 Mbps

- The HT PHY operates at both 2.4 GHz and 5 GHz and supports data rates up to 600 Mbps in the standard, though production devices only support up to 450 Mbps.

- The VHT PHY operates only in the 5 GHz band and supports a maximum data rate of 6933.3 Mbps.

- 802.11i introduced security enhancements to the 802.11 standard.

- 802.11e introduced QoS enhancements to the 802.11 standard.

- 802.11u introduced interworking with external networks to the 802.11 standard.

Review Questions

1. What is the maximum data rate of an HR/DSSS PHY transmission?
    a. 1
    b. 2
    c. 5.5
    d. 11

2. In what frequency band does the VHT PHY operate?
    a. 2.4 GHz
    b. 5 GHz
    c. 60 GHz
    d. 700 MHz

3. What PHY is designed to use available frequencies in the television space?
    a. VHT
    b. S1G
    c. TVHT
    d. HT

4. What PHY is likely to be most used by low output power, long-range devices that require long battery life?
    a. DSSS
    b. HR/DSSS
    c. S1G
    d. VHT

5. What amendment, which is to be ratified after 802.11-2016, will be used by IoT devices?
    a. 802.11ad
    b. 802.11ac
    c. 802.11n
    d. 802.11ah

6. What is the maximum spatial streams supported by an 802.11 n (HT) PHY?
    a. 1
    b. 2
    c. 3
    d. 4

7. What is the maximum channel width supported by an 802.11ac (VHT) PHY?
    a. 20 MHz
    b. 40 MHz
    c. 16 MHz
    d. 160 MHz

8. What amendment to the 802.11 standard provided for enhanced security to resolve the weakness in WEP?
    a. 802.11i
    b. 802.11e
    c. 802.11u
    d. 802.11r

9. Why should only 20-MHz channels be used in 2.4 GHz WLANs?
    a. The available frequencies in the 2.4 GHz band are too few to support wider channels.
    b. No PHYs supports channels wider than 20 MHz in the 2.4 GHz band.
    c. Because they are more compatible with the 22-MHz channel widths used in older PHYs.
    d. Because 2.4 GHz WLANs use higher output power than 5 GHz WLANs.

10. What amendment to the 802.11 standard provided for QoS in WLANs?
    a. 802.11i
    b. 802.11e
    c. 802.11u
    d. 802.11r

Review Answers

1. **D is correct.** HR/DSSS can support data rates of 1, 2, 5.5 or 11 Mbps.

2. **B is correct.** VHT (802.11ac) operates only in the 5 GHz band.

3. **C is correct.** TVHT implements 802.11 in TV white spaces, which are frequency ranges unused by local television broadcasts.

4. **C is correct.** S1G-based devices will support long-range communications because of the longer wavelengths used. They can also use lower output power settings and extend battery life.

5. **D is correct.** 802.11ah is designed and optimized for use in IoT devices among other such devices.

6. **D is correct.** HT (802.11n) supports up to four spatial streams, though most devices in production support three or fewer.

7. **D is correct.** 802.11ac (VHT) can support up to a 160-MHz channel, though they should not be used in any network at this time. A WLAN AP would only very rarely need to be configured with channels wider than 40 MHz.

8. **A is correct.** 802.11i introduces TKIP/RC4 (WPA) and CCMP/AES (WPA2)

9. **A is correct.** The 2.4 GHz band lacks the frequency space to implement channels wider than 20 MHz today effectively.

10. **B is correct.** 802.11e introduces QoS for WLANs.

Chapter 4 – Channel Plans and Performance Factors

Objectives Covered

2.3 Select appropriate channels
    2.3.1   Channel selection best practices
    2.3.2   Common channel selection mistakes

2.4 Identify factors impacting wireless LAN (WLAN) performance
    2.4.1   Coverage requirements
    2.4.2   Capacity requirements
    2.4.3   Required features
    2.4.4   Poor configuration and implementation

3.6 Understand the basic requirements for voice over WLAN (VoWLAN)
    3.6.1   Latency
    3.6.2   Jitter
    3.6.3   Signal strength

A primary factor that impacts WLAN performance is channel selection. When implementing a single AP, channel selection is easy; when implementing larger networks with multiple APs, the selection process becomes more complex. This chapter provides an overview of channel planning and the factors that impact WLAN performance. Such knowledge is important for the CWS professional to make proper recommendations for the acquisition of WLAN hardware and software. Additionally, for the CWS professional within an organization, this knowledge is required to make effective decisions for WLAN hardware and software purchases.

WLAN Performance Factors

To accomplish a well-performing WLAN implementation, the demands of coverage and capacity must be met. Additionally, feature requirements should be considered to ensure that the needed features are available in the final installation. Finally, configuration and implementation must be considered from a management perspective.

Coverage

The first specification that must be met in any WLAN deployment is coverage. If there is no signal, the client cannot connect. For this reason, it is important to understand the basic factors that impact coverage. These factors can be summarized as follows:

- Radiation Patterns
- Output Power
- RF Behaviors

These have all been covered to some extent in preceding chapters. Now, specific scenarios will be explored to help you understand how coverage may be accomplished. Before exploring the specific scenarios, the term coverage must be properly defined.

Consider the process of painting a wall. You want the paint to cover the entire wall. However, the coverage you desire is not simply some paint in all areas; instead, you want the right amount of paint in all areas. In fact, it is not uncommon for a painter to say, "That area is not covered properly. Paint it again." You can take this further and consider painting a light colored paint on a dark wall. In this case, multiple coats would be required to "cover" the wall properly. The same holds true for WLANs. You do not simply want some signal at required locations; you want the right signal at required locations. For this reason, when planning coverage, you must determine the minimum signal strength that is acceptable for coverage areas. Some areas may allow for different minimum signal strengths.

For example, it is not uncommon to determine that VoIP deployments require signal strengths of -65 dBm or -67 dBm. If -65 dBm is your target, you must ensure that the coverage is -65 dBm or better in the intended area. Like the painter, some signal is not enough. Coverage must be achieved, and coverage is defined by signal strength and not simply the ability to say, "You can see the wireless network from there."

> *Coverage* is defined as the proper signal strength at target locations. This signal strength is typically specified in dBm. For example, the proper signal strength may be defined as -70 dBm or stronger. Coverage is achieved when the specification is met.

Standard Office Space

The standard office space is one with carpeted floors, separated rooms and often large spaces with office dividers. Coverage in such environments is easily achieved today, and WLAN designers and engineers are typically able to accomplish it with standard best practices and modern wireless network design software, such as Ekahau Site Survey, iBwave Wi-Fi, TamoSoft Survey and AirMagnet Survey Pro. However, to achieve proper coverage as a CWS, it is important to learn the basic issues faced and solutions provided in these office spaces.

In office spaces, some common trouble areas are elevator shafts, stairwells, and restrooms. The building materials used in these spaces cause more RF absorption and reflections and result in an increased signal loss. They must be considered during deployment to ensure proper coverage. For example, placing APs closer to restrooms and stairwells can help to provide coverage there. The same is true for elevators, but you must also remember that they move up and down. If coverage must exist during the elevator ride, it must be remembered that the passengers will move from floor-to-floor and, roaming must be fast for connections to be maintained.

> It is important to know that some areas have building codes that do not allow mounting of devices in stairwells or elevators. This information must be considered by the design engineers and should be known by a practicing CWS.

After considering problem areas, covering the general office space is really just a factor of AP placement and output power settings. As you explore channel planning in more detail later in this chapter, you will see that the strategic use of channels is key. The goal is to have as little co-channel interference (CCI) as possible while understanding that some CCI will exist in most deployments. CCI occurs when a WLAN device can hear RF signals from a Basic Service Set (BSS) to which it is not connected.

> Remember that a BSS is an AP and the clients it serves. An AP may implement more than one BSS by implementing multiple SSIDs, but all implemented BSS networks that share a radio will operate on the same channel.

In actual deployments, coverage cannot be considered alone, but it must be considered in concert with capacity. You must have coverage everywhere you require access, but that coverage must also provide sufficient capacity for the use cases in that area. Therefore, conference rooms and break areas where people may congregate as well as standard office spaces must be specially considered when planning coverage and capacity.

Figure 4.1 shows a coverage plan for a typical office space. Notice that all internal areas are covered with a green or green/yellow color. The settings in the Ekahau Site Survey (the tool used to generate the image) suggest that this indicates that coverage is better than -70 dBm everywhere internally. This is the view from the 5

GHz band. Each AP is set to 40 mW of output power. While this plan provides coverage, it cannot be assumed that it would also provide capacity. You will learn how to ensure that capacity is provided in the later sub-section: *Capacity*.

Warehouse

A warehouse introduces new concerns for coverage. In many cases, metal shelving causes excessive reflections, and the dense items stored on the shelves may cause significant absorption. For these reasons, it is not uncommon to require many more APs to cover all of the work areas, as depicted in Figure 4.2.

Notice the many APs that are required within the shelving areas to provide sufficient coverage. By comparison, notice the few APs that are required to provide coverage in the more open spaces of the warehouse. For perspective, the floor plan in Figure 4.2 is approximately 300 meters.

Figure 4.1: Coverage Plan for an Office (Generated using Ekahau Site Survey)

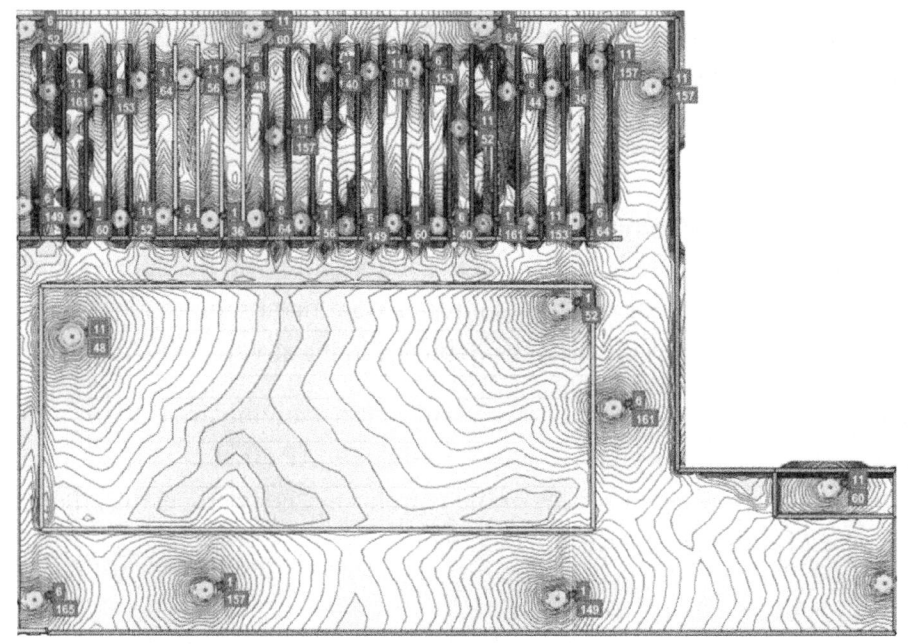

Figure 4.2: Coverage Plan for a Warehouse (Generated using Ekahau Site Survey)

In Figure 4.2, APs with standard omnidirectional antennas are used. However, in such deployments, it is not uncommon to use directional antennas that direct the energy down the hallways between shelves. At times, this alternate design can help to eliminate some co-channel interference. An important consideration when deploying this many APs is wired network provisioning and power provisioning. Often Power over Ethernet (PoE) connections will be used to provide both network connectivity and power.

> *PoE* uses Ethernet cables to provide both power and wired network data connectivity to APs. It uses the unused wires in the cable to provide the power. In most cases, up to 100-meter runs can be implemented without power or data problems.

Outdoor Areas

Outdoor areas present a unique challenge because coverage must be provided, in many cases, over large spaces. With fewer reflective objects, this means that more APs may be required to provide the right balance between coverage and capacity. However, it is also important to remember that outdoor spaces may not be as saturated with users. Determining user counts is an essential part of planning a deployment, including outdoor areas.

Figure 4.3 shows an outdoor deployment. For perspective, the width of this outdoor coverage area is 61 meters. Two semi-directional antenna deployments cover the area very well. There are 12 picnic tables with seating space for eight people at each; therefore, these two dual-band APs should suffice for coverage and capacity. The use of semi-directional antennas helps to prevent excessive interference with the indoor network as well, though, in this scenario, only two APs are required indoors, and the two APs on channel 11 at 2.4 GHz are sufficiently separated by space as well as 10-dB attenuating outdoor walls and indoor drywall barriers. However, each scenario is unique, and the CWS must consider the potential issues when determining the capabilities of varied WLAN solutions.

Additional issues with outdoor areas are AP mounting locations and power and network provisioning. A backhaul to the network can be wired, or it can use an outdoor mesh network; however, this solution does not resolve the power provisioning problem. In each case, the determination of available power sources and how power will be provisioned at the mounting location must be determined.

> *Mesh networks* are built using links to other APs to gain an eventual connection to the wired network. For example, one AP may pass through another that is itself connected to an AP with wired access. Throughput issues must be considered in such deployments.

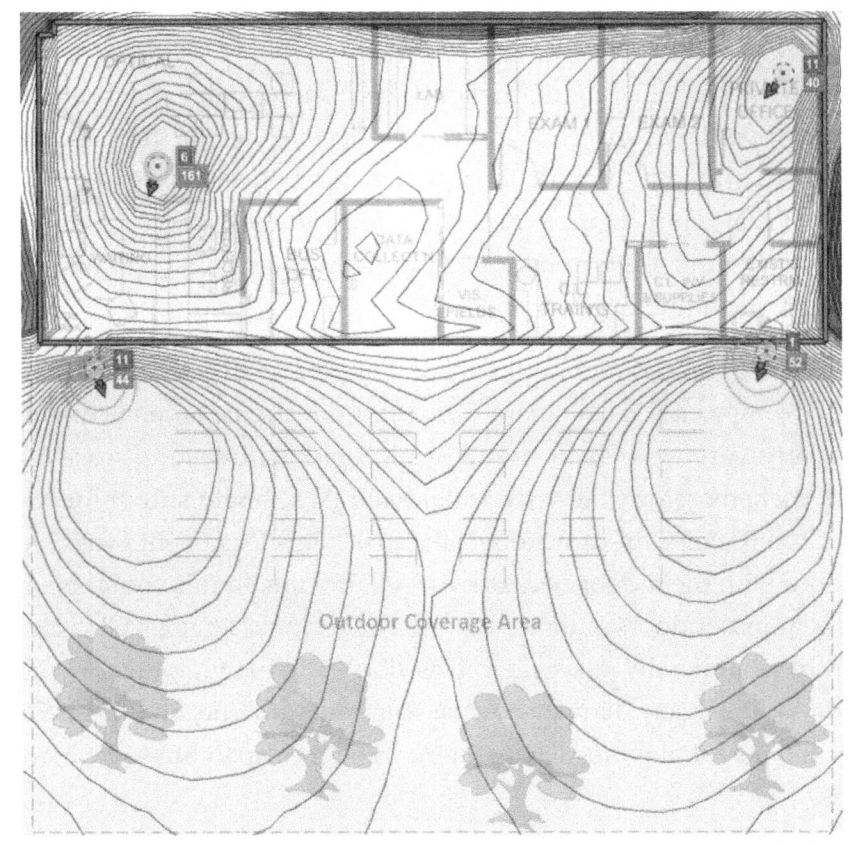

Figure 4.3: Coverage Plan for Outdoor Area (Generated using Ekahau Site Survey)

Capacity

Capacity is the ability of the network to achieve a particular load of client devices and applications used by those devices. Capacity is a factor of the number of devices as well as the device types and application types used. Without this knowledge or at least an estimation, the WLAN sales professional or project manager cannot adequately predict capacity demands.

> Capacity is a factor of devices, device types and applications used. Gathering this information is essential when determining the capacity demands of a WLAN. Without this information, it is impossible to recommend a WLAN solution to meet a customer's needs.

Device Counts

When considering device counts, do not make the error of counting users. Many users will carry three or more devices, and all devices must be factored into the device count. Additionally, the device type is significant. The issue is not whether it is a mobile phone or laptop or some other device, but the issue is the capabilities of the device. A single stream device will consume the AP resource for some window of time regardless of the amount of data it can send in that window. Therefore, devices with lower supported data rates will consume more of the AP's time to send the same amount of data than devices with higher supported data rates. Table 4.1 shows the essential details that should be tracked for known client devices.

Applications

In addition to the devices themselves, you must know the applications they will run. Some devices only run a single application (e.g., barcode scanners), and others run several application types. Table 4.1 shows general applications as well. When creating an application inventory, the requirements of the application should be documented. Areas to consider include the following:

- **Throughput requirements:** each application should be defined in relation to its throughput requirements, typically in Kbps (kilobits per second).

- **Utilization:** each application should be defined in relation to its activity or utilization. Stated simply, what percentage of time is the application communicating on the network?

- **Latency and jitter requirements:** latency is the amount of delay incurred when sending data from its source to its destination. Latency can refer to either one-way latency (the time to travel to the destination) or round-trip latency (the time to travel to the destination and receive an acknowledgment from the destination). Applications like VoIP often require less than 150 milliseconds (ms) of one-way latency. Jitter is the variance in the reception of streamed data at the receiver (destination). If one packet arrives in 50 ms and the next in 65 ms, 15 ms of jitter is incurred. Some solutions use buffers to resolve this issue.

Device Description	PHY Supported	Streams	Data Rate	Applications
iPhone 7 (75)	802.11ac/802.11n	2x2	400/144.4	Web, Messaging, VoIP
iPhone 6 (35)	802.11ac/802.11n	1x1	200/72.2	Web, Messaging, VoIP
Nexus 6x (40)	802.11ac/802.11n	2x2	400/144.4	Web, Messaging, VoIP
ASUS Laptop (23)	802.11ac/802.11n	3x3	600/216.7	Database, Web, Messaging, File Transfer
MacBook Pro (17)	802.11ac/802.11n	3x3	600/216.7	Web, Messaging, Very Large File Transfer
Desktops (5)	802.11ac/802.11n	4x4	800/216.7	Web, Messaging, Very Large File Transfer, Real-Time Monitoring

Table 4.1: Device Inventory for Wi-Fi Planning (Data Rates Assume a 40-MHz Channel (Max) at 5 GHz and a 20-MHz Channel (Max) at 2.4 GHz)

With this information, you can determine the demands that the application will place on the network. Remember that each device may run one or several such applications. Ultimately, you should be able to define the throughput requirements of the applications in Kbps or Mbps. Real-time applications (e.g., VoIP) will also demand latency and jitter constraints.

> In networking, *bps* stands for bits per second. *Kbps* is kilobits per second or 1000 bps. *Mbps* is megabits per second or 1,000,000 (one million) bps. Finally, *Gbps* is gigabits per second or 1,000,000,000 (one billion) bps.

Capacity can often be increased by adding more APs to allow for fewer client STAs per AP, disabling lower data rates to improve airtime usage by transmitting data faster, using as few SSIDs as possible, and removing APs in a badly over-engineered WLAN with excessive CCI.

Feature Requirements

Several vendor-specific features affect the performance of WLANs:

- **Band Steering:** The ability to direct clients to 5 GHz channels is called *band steering*. This feature may improve performance if it results in more clients (stations or STAs) connecting to 5 GHz networks instead of 2.4 GHz networks. The 2.4 GHz band is more congested and offers fewer channels (as you will see later), which is why the 5 GHz band is preferred.

- **Load Balancing:** The ability to distribute the load of clients among APs in a more balanced way can improve performance overall. This feature is called *load balancing* and is implemented through the configuration of maximum allowed client connections, or it can be an explicit feature of an AP.

- **Auto-Channel Selection:** Sometimes called Radio Resource Management (RRM) or Automatic Resource Management (ARM), *auto-channel selection* allows the AP or controller to select the best channel and power settings based on RF activity in the area. When configured correctly, the feature may improve performance. When misconfigured, it can wreak havoc on the network.

- **Network Monitoring:** Implemented in several different ways, *network monitoring* allows the APs, controllers or management system to report on the health of the network. Reported information may include connection statistics, throughput measurements, spectrum analysis details and more. Network monitoring provides administrators the knowledge they require to tweak and tune the WLAN.

Additional factors that affect performance are available in nearly all vendor solutions (vendor-neutral):

- **Output Power Settings:** The output power settings of an AP are critical when improving the performance of a WLAN implementation. If the output power setting is too high, performance will suffer. If the output power setting is too low, the needed coverage will not be provided. In general, you should configure output power settings on the AP at the same level as the clients, but this can be a significant challenge as clients vary significantly in output power settings.

- **Compatibility Settings:** The settings that allow for older client STAs to connect to the AP or that disallow them can be grouped as compatibility settings. They include settings such as support for older PHYs and support for non-MCS data rates (pre-802.11n data rates).

- **Data Rate Management:** Disabling low data rates may help improve a WLAN's performance; however, doing this can also remove support for some older clients. In addition, some clients will remain connected to the AP when they should roam to another AP and continually attempt to send data at data rates that can no longer be supported based on reduced signal strength.

The CWS should understand all of these vendor-specific and vendor-neutral items.

> Remember the basic benefits of band steering, load balancing, auto-channel selection, network monitoring, output power settings, compatibility settings and data rate management for exam day.

Configuration and Implementation

Poor configuration and implementation are significant factors that can result in poorly performing WLANs. To avoid this dilemma, be sure to follow these guidelines:

- Understand vendor best practices for configuration and installation.

- Hire or assign a well-trained engineer (e.g., a CWNE) to perform the network design.

- Avoid over-engineering the network (i.e., installing too many APs).

- Avoid under-engineering the network (i.e., installing too few APs).

- Ensure that the implementation meets both coverage and capacity requirements.

- When using auto-channel selection, validate the channels used in the installation.

- When resolving coverage issues, do not rely on increased output power (over the design specification) as the primary solution. Instead, move APs or add more APs to resolve the issue.

> Increasing the output power may resolve coverage issues, but it may introduce significant interference issues that diminish performance in other key areas of the network. This is why moving or adding APs is typically preferred.

Channel Selection

As you learned in the preceding section, channel selection is a key factor in achieving capacity. In this section, you will learn about the available channels in 2.4 GHz and 5 GHz implementations and the constraints related to channel selection. You will also learn about channel selection best practices and common mistakes.

802.11 Channels

Channelization, which is merely a fancy term for how channels work in a network, of 802.11 networks varies by several factors. First, the IEEE standard defines generally available channels. Second, regulatory domains and the agencies that manage RF communications constrain the available channels. Third, channel usage may be restricted in specific areas due to other RF communications, such as radar.

In this section and subsections, the available channels for use with 2.4 GHz and 5 GHz are explained. You will learn about the available channels, channel widths and issues related to channel selection.

2.4 GHz Channels

In the 2.4 GHz band, 14 channels are specified in the 802.11 standard. Most regulatory domains allow for only 11 of these channels (channels 1-11). It is vital for you to know the channels that are available in the regulatory domain in which the network will be deployed, but you are not required to have this knowledge for the CWS exam since it is region-specific.

Figure 4.4 shows the 14 channels that are available in the 2.4 GHz band. The representation of the channels is intended to show how they overlap. For example, channels 2-4 significantly overlap with channel 1, and channels 3-5 significantly overlap with channel 6. For this reason, the typical recommendation is to use only channels 1, 6 and 11 in the 2.4 GHz band. Sadly, this recommendation is either not always known or not always followed, which leads to abysmal performance for 2.4 GHz WLANs.

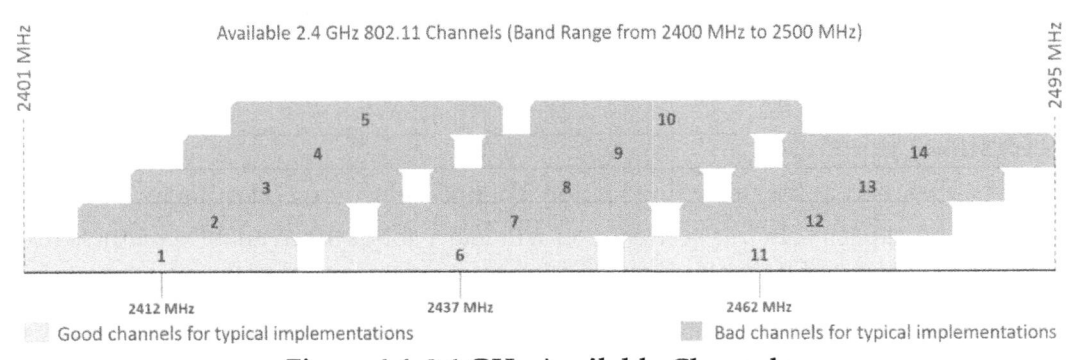

Figure 4.4: 2.4 GHz Available Channels

Figure 4.5 shows WLAN activity in a residential area in the 2.4 GHz band. Notice that networks have been detected on channels 2, 3 and 8. Additionally, the AP on channel 2 also uses a 40-MHz channel, which makes matters far worse. This figure illustrates the poor results that are often achieved with auto-channel selection, which is most often used in these residential deployments. Sadly, many enterprise deployments suffer from the same problem.

> Remember that channels 1, 6 and 11 are the best choice for channels in the 2.4 GHz band for nearly all network implementations. In some uncommon scenarios, other channels may be used, but this is considered the best practice.

Other than available channels, regulatory domains do not have restrictions related to radar detection in the 2.4 GHz band. However, local regulatory agencies will constrain output power levels in this band as well as the 5 GHz band.

Finally, 802.11n (HT) introduced the allowance for 40-MHz channels in both the 2.4 GHz and 5 GHz bands. Before this, either 20-MHz or 22-MHz channels were used. 20-MHz channels were used for the newer 802.11g (ERP) PHY devices, and 22-MHz channels were used for the older 802.11 (DSSS) and 802.11b (HR/DSSS) PHYs. However, even though they are available, 40-MHz channels should not be used in the 2.4 GHz band due to frequency bandwidth constraints. Figure 4.6 shows the impact of a 40-MHz channel on the 2.4 GHz band. In a clean environment devoid of other 2.4 GHz communications, you may be able to use a single 40-MHz channel. Given that so few such environments exist, it is just impractical.

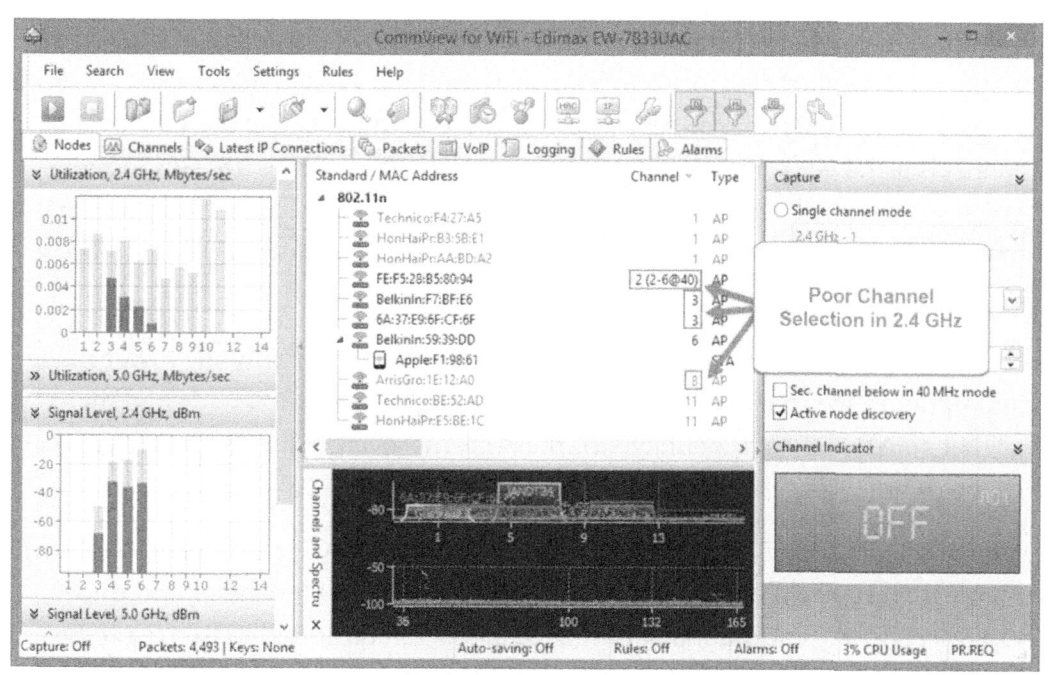

Figure 4.5: Poor Channel Selection (Reported in CommView for Wi-Fi)

5 GHz Channels

The 5 GHz band has many more 20-MHz channels than the 2.4 GHz band. In fact, it has 25 channels in regulatory domains that support them and when the WLAN devices support all of the channels. Even in the more restrictive regulatory domains, far more than the three useful channels in the 2.4 GHz band are available.

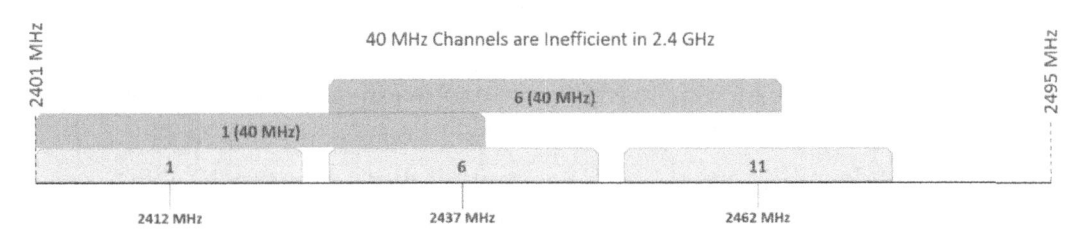

Figure 4.6: 40 MHz Channels and the Impact in 2.4 GHz

The lower range depicted in Figure 4.7 as well as the upper range is supported throughout most of the world. This provides at least 12 20-MHz channels that can be used in the 5 GHz band. As you can see, this is much better than the available channels at 2.4 GHz. Within the United States, all 25 channels are supported; however, the middle range channels are considered DFS channels and must monitor for radar and cease use if radar is detected. For most implementations, radar will not be a problem.

However, it is essential to know that many clients support only the lower range and upper range channels. When this is the case, it is essential to design the WLAN so that an AP is always available for these clients, even if the middle range channels are used.

Unlike 2.4 GHz, the 5 GHz band can easily support 40-MHz channels. As shown in Figure 4.7, six 40-MHz channels are available using only the lower and upper ranges. Five or six additional 40-MHz channels can be added to that list if the middle range can be used. Only five additional channels will be available if channel 144 is not supported. When channel 144 is supported, six 40-MHz channels are added for a total of 12 40-MHz channels.

While 802.11ac introduced support for 80-MHz channels, they are not recommended for most deployments. This recommendation is based on the fact that only six 80-MHz channels are available, even if all ranges may be used. Additionally, 160-MHz channels are rarely used. The FCC in the United States defined two 160-MHz channels, as depicted in Figure 4.7. However, using the 80+80 support in 802.11ac, a third may be acquired, assuming all ranges are available. If you use all three channels (two 160 MHz and one 80+80 MHz), you will be dealing with similar issues of frequency reuse experienced with 2.4 GHz. Good frequency reuse results in as little interference from the same channel (co-channel interference) as possible.

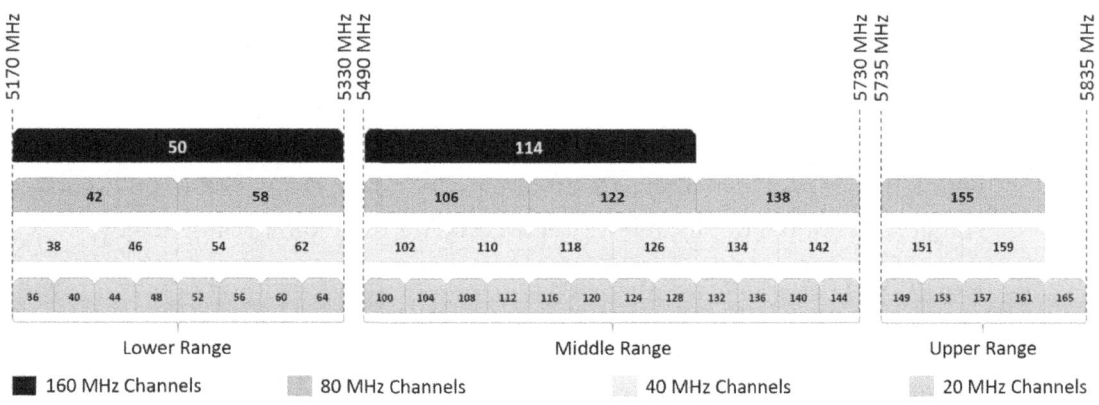

Figure 4.7: 5 GHz Available Channels

> *Frequency reuse* is a reference to the intentional and strategic reuse of channels in a wireless network design. For example, if you place an AP on channel 36, at some point in a large deployment, you will need channel 36 again.

A simple way to remember recommendations for channel width selection at 5 GHz is to note the colors in Figure 4.7. Green (20 MHz) is a very safe choice. Yellow (40 MHz) should be used with caution. Red (80 MHz) should cause you to stop and think very seriously before deploying it. Black (160 MHz) means that it should simply not be used until more frequency space is made available.

> High-density WLANs (e.g., stadiums, conference centers) operating at 5 GHz most often use 20-MHz channels. This decision allows for the deployment of more APs in the coverage area, which increases capacity. Today, standard office spaces often use 40-MHz channels.

Channel Selection Best Practices

With an understanding of the available channels associated with 2.4 GHz and 5 GHz, you can better understand the following channel selection best practices:

- Choose a channel that has the least RF activity in the target coverage area. This activity includes other 802.11 WLANs and non-802.11 RF devices, such as video cameras, microwave ovens and more.

- Channels must be staggered throughout a deployment to enable frequency reuse. For example, at 2.4 GHz, you can alternate among channels 1, 6 and 11, and in 5 GHz, you can alternate among many more. Figure 4.8 depicts such a staggered channel deployment prepared using the Ekahau Site Survey.

- Use 20-MHz channels for all 2.4 GHz deployments.

- Only use 40-MHz channels for standard office deployments with 5 GHz when the clients require extra momentary throughput; otherwise, use 20-MHz channels with 5 GHz as well.

- Avoid 80-MHz channels as much as possible.

- Do not use 160-MHz channels in any enterprise deployment.

- Use DFS channels but provide for clients that do not support them.

> *Dynamic Frequency Selection (DFS)* defines channels that must be monitored for radar activity. If radar activity is detected, the channel must be avoided by a WLAN. Most such channels are in the middle range defined in the later subsection: *5 GHz Channels*.

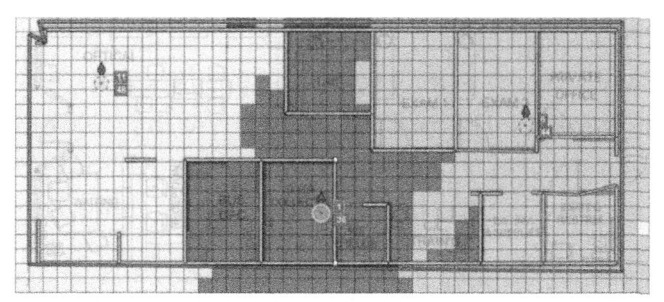

Figure 4.8: Channel Staggering in 2.4 GHz and 5 GHz (Shown in Ekahau Site Survey)

Channel Selection Mistakes

The most common channel selection mistakes are often the opposite of recommended best practices:

- Choosing channels without defining existing activity in the target coverage area.

- Using the same channel throughout the deployment (with the exception of a WLAN architecture known as Single Channel Architecture [SCA]).

- Using 40-MHz channels at 2.4 GHz.

- Using 40-MHz channels at 5 GHz without determining whether clients require their use.

- Using 80-MHz channels at 5 GHz when it is not required.

- Using 160 MHz channels – ever.

Voice over WLAN (VoWLAN) Requirements

One of the most demanding use cases of WLAN networks is Voice over IP (VoIP) running on the WLAN. Such use cases are typically called Voice over WLAN or VoWLAN. In this section, you will review VoIP communication basics and the factors that impact VoWLAN implementations.

Voice Communication Basics

A CWS candidate must ask, "How do we get voice communications to travel over data networks?" The answer is simple: convert the audible telephone conversation into digital data packets. This conversion is accomplished by encoding standards, and the voice data packets are then transferred using standard or proprietary voice communication protocols. This is really no different than any other data communications process in terms of its technical nature; however, voice data packets do come with demands that are not seen in traditional data packets.

For example, if you are sending a file to a server using FTP, it does not matter if a few packets arrive out of order or if there happens to be a delay between the arrival of one packet and the arrival of the next packet that is less than a particular threshold. Of course, if there is an extended delay, it will slow the communications down, but the data will eventually arrive at the destination, and the receiver will reassemble the packets in the appropriate order. Voice traffic will not tolerate packets that are significantly out of order or with high variance in packet delivery times. If there is an extended delay, the call will be dropped, or the call quality will suffer.

There are humans at both ends of a Voice over IP (VoIP) communications link. They both talk and listen, and they have expectations that have been set by the analog telephone networks of the past. If they do not hear any sound for some variable length of time, they will assume that the call has been dropped or the person on the other end has disconnected. If the sound quality is inferior, particularly to the point where they cannot understand one another, they may give up on the conversation. There are expectations of quality that must be met with VoIP data that have not traditionally been required of other data types. In fact, we often refer to "carrier grade" or "carrier quality" VoIP communications. This term means that we have accomplished a quality of sound and communication speed that is at least equivalent to the traditional Public Switched Telephone Network (PSTN).

Since we are transmitting the VoIP packets over the same physical network as traditional data packets (e.g., e-mail, database access, file transfer, printing), we are layering voice over the data network. We are using the same network devices, cables, and software used for traditional data to transfer voice data. This layering places a new demand on the network. The demand is that the data network must be able to differentiate between various packet types and give priority to voice data so that the quality expectations of the VoIP users are met. This technology is called Quality of Service (QoS). QoS provides queue management and prioritization of traffic to allow the most important traffic to travel through first.

Because voice traffic must move at a rapid speed across the network and because it would not provide a benefit to resend the traffic if it is corrupted or lost (given it is intended to be a stream of data processed as it arrives), we use the User Datagram Protocol (UDP) to send most VoIP data packets. UDP is a connectionless protocol, unlike the Transmission Control Protocol (TCP). TCP has far too much overhead to transmit voice packets as rapidly as they require.

You may wonder why there is no benefit associated with resending corrupted or lost voice packets. The reason is simple. Think about how long it takes you to say the word *don't*. It will take you far less than a second. Now, imagine you are having a conversation on a VoIP phone, and you say the following sentences, "Don't push the button. Pull the lever." Further, imagine that the word *don't* was lost in transmission, and the system decided to resend it. Because of the sequencing problem, the user on the other end hears the following, "Push the button. Don't pull the lever." This reordering could theoretically happen because the phrase "Push the button" made it through, while the word *don't* did not make it through. When the word *don't* was retransmitted, it was placed before the phrase "Pull the lever". The result is the complete opposite of the intended message. Do you see why retransmitting lost audio packets would be useless and possibly damaging?

> In real-world VoIP deployments, entire words are not typically rearranged like this. Yet, the example helps to illustrate the importance of processing data as it arrives in voice communications.

Instead of rearranging the words, in most actual VoIP deployments, the listener would just not hear the word *don't*; however, the reality is that it is more complex. The listener would probably hear something like, "D---t pu—the ---ton. Pu-- --- -- ver." All the dashes represent either sounds that are unintelligible or complete silence. The point is that the network does not usually drop exact words but rather portions of audio much less than a complete word, resulting in what we usually call a "bad connection."

The process used to convert voice to data is really a five-step process. The steps are as follows:

1) Convert the sound waves to analog electrical signals.
2) Convert the analog electrical signals to digital signals.
3) If required, compress the digital signal.
4) Create packets from the signal data.
5) Transmit the packets on the data network.

> Many VoIP implementations use a 10-ms sampling window. The device creates a new audio sample every 10 ms. 10 ms is 1/100 of a second, so single words require multiple samples or multiple packets. The device typically transmits two or more audio samples in each packet.

For this process to work correctly, several metrics must be considered, and they will be covered in the next section: *Factors Impacting Voice Communications on WLANs*.

VoIP devices use something called a *CODEC* (coder/decoder) to prepare audio samples for delivery. Two common codecs are G.711 and G.729. Table 4.2 lists specifications for an example implementation of these two codecs. The bandwidth is that which is required in throughput to transmit the data to its destination. The sample size is the total size of each voice data payload in the implementation. The MOS is the mean opinion score. A higher number indicates better call quality. The compression column indicates whether the data is compressed or not; when comparing G.711 (no compression) with G.729 (compression), you can see the significant impact compression has on required throughput.

Based on the data in Table 4.2, which does not consider overhead added by the UDP protocol, you can see that a single call will require several bytes of data per second with a need for 64 Kbps with G.711 and 8 Kbps with G.729. The required throughput is only part of the puzzle. Additionally, the packets must be delivered in a timely and consistent manner (with low latency and low jitter levels), which is discussed next.

> You will not be required to memorize codec specifications for the CWS exam; however, you should know that different codecs result in different requirements. You should also know that the codec usually determines the sampling size in a given implementation.

Codec	Bandwidth (kbps)	Sample Size (bytes)	MOS	Compression	Packets Per Sec
G.711	64	240	4.1	No	33
G.711	64	160	4.1	No	50
G.729	8	40	3.92	Yes	25
G.729	8	20	3.92	Yes	50

Table 4.2: Sample Implementation Metrics for G.711 and G.729

When selecting a VoIP codec, the following should be considered:

- **Device support:** All devices on the network must support the codec, or you must provide a gateway for conversion between the various devices. Using gateways throughout your network for this purpose can cause latency and certainly increases cost. It is usually best to choose a codec that is supported by all of your hardware. For example, G.711 and G.729 are supported by all Cisco voice equipment.

- **Network activity:** You must consider the activity on the network before introducing VoIP. If the network is already busy, the VoIP traffic will only add to the current load. A codec with compression may be a good choice in congested networks that cannot be upgraded. The preference, of course, is to upgrade the network.

- **Network packet loss:** Some networks lose packets even though the entire network is not congested or busy. Packet loss can happen because a network choke point exists (a point where the different network segments converge, such as a single router connecting all segments). In this case, you can use a lower bandwidth codec, but, again, the preference is to eradicate the choke point through hardware upgrades or replacements.

- **Compression:** Using a codec that implements compression reduces the size of sample data and therefore the size of packets. If the quality is sufficient with compression, it is often best to use a codec that offers it.

- **Sample sizes:** You must consider the sample size used by the codec. In addition to compression, the sample size impacts the packet size and the number of packets needed to transmit one second of voice data. Networks with higher packet loss may be better configured with a codec using more, but smaller, packets.

> Remember these factors for exam day. You may see questions related to these important considerations for WLANs that will support VoIP. Understanding the factors of device support, network activity, network packet loss, compression, and sample sizes is important.

Factors Impacting Voice Communications on WLANs

WLANs introduce issues to voice communications that are not familiar problems on wired networks. First, they can increase latency. Second, they often cause more jitter in communications. Third, they require more signal strength for the clients than most standard data applications so that higher data rates can be achieved. This section explores these three issues in greater depth.

Latency on WLANs

Unlike wired networks, wireless networks require that the transmitting STA wait until the medium is clear to transmit. This means that the STA does not have uninterrupted continual access to the network. This waiting procedure results in greater latency on WLANs than on wired LANs and can affect VoIP communications.

Because of this, the IEEE developed QoS options in the 802.11e amendment that are now part of 802.11-2016. The QoS in 802.11 networks is a probabilistic QoS for the wireless link and does not guarantee priority. It increases the likelihood that a high-priority communication will gain access to the medium before communications with lower priority. The QoS should be enabled on all WLANs, but it is particularly important on WLANs supporting VoWLAN communications.

Jitter

As stated previously, jitter is the variance in delay or latency. This variation in delivery times can cause serious problems in the quality of voice communications. To address the issue of jitter, many VoIP receivers use a jitter buffer, which is a collection of incoming packets that are passed to the decoder in a more consistent sequence. When dealing with live audio streams, the jitter buffer can only correct moderate jitter problems. Congested WLANs can cause extensive jitter problems.

To address jitter problems in WLANs, consider the following solutions to reduce congestion in the BSS:

- Install more APs
- Remove unnecessary wireless clients (e.g., printers, desktops)
- Remove non-Wi-Fi sources of interference
- Optimize the channel plan to avoid channel over-congestion

Signal Strength Requirements

Finally, to ensure a stable and fast connection on the wireless link, a good signal is required at the location of the VoWLAN client. Vendors vary on this signal strength requirement, but it is typically somewhere between -65 dBm and -70 dBm, and many recommend -67 dBm. When designing coverage for the WLAN that will support VoIP clients, the areas where telephone conversations will occur must have signal coverage at the required level.

Engineers will use tools like iBwave Wi-Fi, Ekahau Site Survey, TamoGraph Survey and AirMagnet Survey Pro to design for coverage that provides a minimum signal strength according to the vendor requirements. After the WLAN installation, a validation survey should be performed to verify the signal strength coverage and the ability to place and maintain calls while moving around the facility.

> Engineers must have the skills to perform such surveys and tests. The CWS candidate is not required to have these skills, but the candidate should be aware of the need for such testing so that the customer can be appropriately informed.

Chapter Summary

In this chapter, you learned about WLAN performance factors, including coverage and capacity. You also learned about the channels available for both 2.4 GHz and 5 GHz as well as best practices in channel selection. Finally, you learned the specific requirements of VoWLAN implementations and the need to address potential concerns related to them. In the next chapter, you will learn about WLAN security concerns.

Points to Remember

Remember the following important points:

- Coverage is important in WLAN deployments and is usually specified as some minimum signal strength requirement.

- Signal strength requirements are typically stated in dB, such as -65 dBm or -70 dBm.

- Capacity is the ability of the WLAN to service the needed number of clients and the applications they use.

- Capacity can be increased by adding more APs with reduced output power settings.

- The 2.4 GHz band should only be used with 20-MHz (802.11g [ERP] and 802.11n [HT]) or 22-MHz (802.11 [DSSS] and 802.11b [HR/DSSS]) channels.

- 40-MHZ channels should not be used at 2.4 GHz.

- 80-MHz channels should be used with great care at 5 GHz.

- 160-MHz channels should not be used at 5 GHz.

- 25 20-MHz channels are potentially available in the 5 GHz band.

- Some channels may require monitoring for radar in the 5 GHz band, and these are called DFS channels.

- VoIP requires low latency and reliability for proper operations.

- WLANs can introduce jitter due to the use of a shared medium (RF).
- Jitter buffers in receivers can help to reduce problems caused by jitter.
- Removing WLAN congestion problems can improve the performance of VoWLAN communications.

Review Questions

1. What metric is typically used to define coverage requirements?
   a. Latency
   b. Jitter
   c. dBm
   d. Watts

2. Which of the following is a method used to increase capacity?
   a. Add more APs
   b. Use only channel 1 in 2.4 GHz
   c. Increase the output power of APs
   d. Enable lower data rates

3. Which of the following is a 2.4 GHz channel?
   a. 11
   b. 44
   c. 48
   d. 157

4. What 5 GHz channels may require monitoring for radar signals?
   a. DFS channels
   b. All channels
   c. Upper range channels
   d. Lower range channels

5. What channel width should not be used in the 2.4 GHz band?
   a. 20 MHz
   b. 22 MHz
   c. 40 MHz
   d. None of these

6. What channel width is typically used in high-density 5 GHz WLANs?
   a. 20 MHz
   b. 40 MHz
   c. 80 MHz
   d. 160 MHz

7. What environment is most likely to use 40-MHz channels in the 5 GHz band?
   a. Stadium
   b. Conference center
   c. Standard office space
   d. An environment requiring support for 802.11b devices

8. How many 20-MHz channels are available in the 5 GHz band based on the 802.11 standard without consideration for regulatory domains?
   a. 11
   b. 14
   c. 24
   d. 25

9. What metric reveals the delay in transmission of a packet from a source to a destination on the network?
   a. Jitter
   b. Latency
   c. Throughput
   d. Contention

10. Which one of the following is a common minimum signal strength recommended for VoWLAN deployments?
    a. -67 dBm
    b. -80 dBm
    c. -83 dBm
    d. -87 dBm

Review Answers

1. **C is correct.** Signal strength in dBm is the typical metric used to define coverage requirements.

2. **A is correct.** Adding more APs can increase capacity when it is done correctly.

3. **A is correct.** 2.4 GHz channels are numbered 1 through 14.

4. **A is correct.** DFS channels must be monitored for radio activity in many regulatory domains.

5. **C is correct.** 40-MHz channels should never be used in the 2.4 GHz band because it provides insufficient frequency bandwidth.

6. **A is correct.** High-density deployments typically use 20-MHz channels, even in the 5 GHz band.

7. **C is correct.** Standard office spaces commonly use 40-MHz channels in the 5 GHz band.

8. **D is correct.** The 5 GHz band includes 25 20-MHz channels, though not all are available in all regulatory domains.

9. **B is correct.** Latency is a measurement of the delay in network communications.

10. **A is correct.** VoIP implementations often recommend signal strengths of -65, -67 or -70 dBm or greater.

Chapter 5 – WLAN Security, BYOD, and Guest Access

Objectives Covered

2.5 Explain the basic differences between WPA and WPA2 security
    2.5.1   Authentication and key management
    2.5.2   Encryption
    2.5.3   Personal vs. Enterprise

3.7 Determine the best solution for BYOD and guest access
    3.7.1   User provisioning
    3.7.2   Captive portals
    3.7.3   Device and software control solutions

Implementing secure WLANs that meet the needs of organizations and employees including BYOD (Bring Your Own Device) and guest access is essential. This chapter introduces the security options that are available in 802.11 WLANs and addresses both BYOD and guest access from the viewpoint of decision makers. You will learn about WLAN security basics, including WPA and WPA2 as well as older security technologies that should no longer be used. Then, you will explore BYOD and guest access.

WLAN Security Basics

To understand the basics of WLAN security, you must first understand why the specific security solutions are used. This section first discusses the need for security in WLANs and then discusses specific standards-based security solutions.

Why is Security Important in WLANs?

WLANs have the same vulnerabilities that are common to all computer networks, and they have vulnerabilities that are specific to 802.11 networks as well. When a technology is implemented in the same way over and over again, it becomes more likely that someone will eventually discover the weaknesses in that technology, if weaknesses exist. This is true for the most popular operating systems and software applications, and it is also true for the most popular wireless local access networks, which are based on the 802.11 standard.

IEEE 802.11 networks showed their biggest security vulnerability when the one real security feature was hacked in the first few years of existence. This security feature was WEP, which will be discussed later in the sub-section *Deprecated Security*. Other optional configuration parameters have been touted as security features over the years, but they were either never intended as such or vulnerabilities were known from their inception, such as MAC filtering and SSID hiding. As you will see, neither of these features provide sufficient security for the network.

The following list of common WLAN vulnerabilities makes it clear that security is needed for implementations:

- **Eavesdropping:** Capturing packets as they traverse the WLAN. Eavesdropping is easily accomplished when encryption is not used. Many packet sniffers (i.e., tools for capturing data packets) are available that capture WLAN traffic.

- **Denial of Service (DoS):** Preventing valid users from using the WLAN through network attacks. A DoS can be implemented at the physical level through RF signal generation or at the network data level through special network attacks. The only real countermeasure to physical level attacks is network monitoring and response. Network data level attacks can sometimes be prevented by implementing features like management frame protection, which prevents inappropriate deauthentication frame communications, for example.

- **Management interface exploits:** Taking advantage of open management interfaces or management interfaces with weak security. Because of this attack vector, default passwords should never be used for management accounts, and account names should be changed when possible.

- **Encryption cracking:** Exploiting weaknesses in encryption algorithms or systems. Early 802.11 encryption solutions were weak. WEP could be cracked in under five minutes within 3-4 years of the initial 802.11-1997 release. Improved encryption solutions, such as those in WPA2, should be used as a countermeasure.

- **Authentication cracking:** Exploiting weaknesses in authentication algorithms. Early 802.11 authentication using Shared Key authentication, which was based on WEP, was very weak. Some later authentication methods (e.g., LEAP) also proved to be vulnerable. The countermeasure is to use strong authentication types, such as EAP-TLS, EAP-TTLS, and properly implemented PEAP.

- **MAC spoofing:** Impersonating a MAC address other than the one in the local NIC (network interface card). This method is used on WLANs to get around MAC filtering or to hide the identity of the attacker's machine in log files. The simplest countermeasure is to avoid the use of MAC filtering as a security solution.

- **Peer-to-peer attacks:** Directly attacking other wireless clients or attacking them through the network. Most peer-to-peer attacks are performed directly against other wireless clients and include file access, malware injection, man-in-the-middle and hijacking attacks. Countermeasures include disabling direct communications among wireless clients, when possible, and implementing good client-based security protection tools, such as client firewalls and special wireless client security solutions.

- **Social engineering**: Using human manipulation to get around technical security solutions. For example, when WPA2-Personal is used, which uses a pre-shared key for authentication, a social engineer may manipulate a user into giving him or her the pre-shared key passphrase. The only significant countermeasure is user training.

This list makes it clear that WLAN security is important. To implement the needed security, you must understand the concepts of authentication, key management, and encryption. These are discussed next. Then, you will explore specific 802.11, standards-based security solutions in WPA and WPA2 certifications from the Wi-Fi Alliance. Finally, you will learn which 802.11 security options are now deprecated and should no longer be used in WLANs.

Authentication, Key Management, and Encryption

In the field of security, the acronym CIA is often used to indicate three important concepts: confidentiality, integrity, and availability. They are sometimes called the CIA security triad, CIA triad or CIA triangle. Figure 5.1 illustrates the concept.

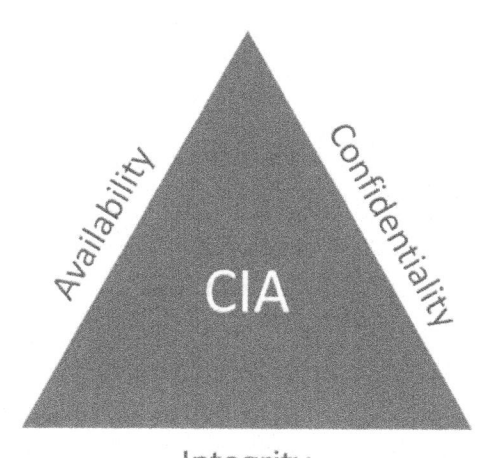

Figure 5.1: CIA Triad

Confidentiality is the concept of keeping private information private. It is accomplished by restricting access to the information when it is stored, transferred or utilized in any other way. During storage, it is achieved by encrypting the data in some cases and restricting access to it in all cases. During the transfer, it is achieved through the use of encryption. During utilization, it is achieved by means of physical security, which controls access to the area where the information is being used on-screen or in print format. An example of weak confidentiality is the WEP encryption in IEEE 802.11.

Integrity is the concept of data consistency. In other words, the data is what it should be. This must be true when the data is transferred from one place to another. For example, a man-in-the-middle attack may involve receiving data from an unsuspecting client and changing it in some way before it is sent on to the destination. When this occurs, the data integrity has been violated. This is usually protected against by using hashing algorithms and CRC methods. A hijacking attack that evolves into a man-in-the-middle attack, then, is an example of an integrity violation.

Availability simply states that the right data is available to the right people at the right time in the right place. This is a factor of data throughput as much as it is of data security. However, if you have provided sufficient throughput, and then an attacker performs a DoS attack, availability suffers. A DoS, then, is an example of a security breach that would violate the principle of availability.

These three concepts must be considered when implementing your WLAN. Not only must you consider how you will provide confidentiality, integrity, and availability, but you must also consider which is most important. For example, stronger encryption may require more overhead that in turn reduces availability (throughput). The same is true for various integrity algorithms. If availability is highly important, you will have to either sacrifice the level of confidentiality and/or integrity or implement hardware that is sufficiently powerful to overcome the overhead. This means that your WLAN will cost more, but it is often worth the cost.

An additional fundamental concept of security is AAA: Authentication, Authorization, and Accounting. Figure 5.2 illustrates the interconnections of the AAA components. Authentication is needed for authorization, and accounting ensures proper actions related to both authentication and authorization.

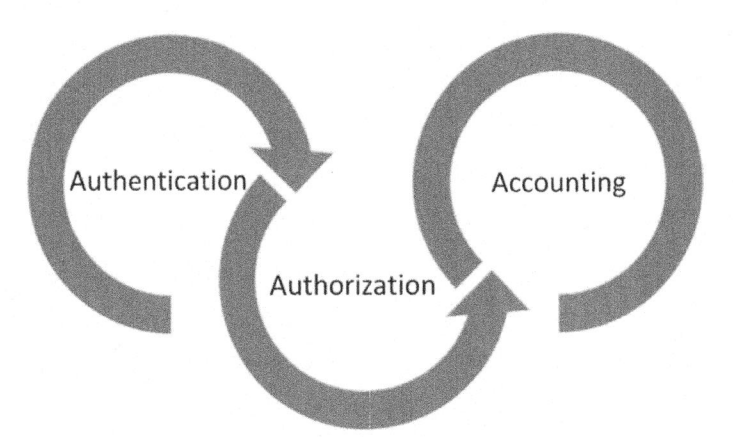

Figure 5.2: AAA Illustrated

In 802.11 networks, authentication is provided primarily through pre-shared keys based on passphrases or an authentication protocol based on the EAP (Extensible Authentication Protocol) model. These are explored in more depth later in this section. Authorization may be implemented through Role-Based Access Control (RBAC), or it may be left to be performed by other network resources (e.g., shares, printers). Accounting is provided, in part, by the APs or controllers on the network and may also be implemented using centralized monitoring solutions. The remainder of this section will focus on authentication, key management, and encryption as they are used in WLANs.

> *Role-Based Access Control (RBAC)* implements authorization using groups or roles. Each connected user or device is placed in a group or role that has a collection of permissions or rights. RBAC makes authorization management simpler than a user-based model.

Authentication is the process used to corroborate or validate the identity of a human or system. It is accomplished using proof-of-identity concepts. Three basic techniques are used alone or in combination:

- Something you know
- Something you have
- Something you are

Authentication with passwords is an example of something you know. Smart cards or certificates are examples of something you have. Biometrics models, fingerprint readers, retina scanners and others are examples of something you are. Many systems use a combination of factors and are called multi-factor authentication systems. For example, inserting a smart card and entering a PIN (personal identification number) is an example of something you have and something you know working together in a multi-factor system.

Most WLANs implement either something you know or something you have. Something you know is used in smaller networks when a passphrase is required to join the WLAN. Something you have is used in larger networks when a certificate is required for authentication. In addition to the certificate, usernames and passwords are often required, resulting in a form of multi-factor authentication.

Do not confuse connecting to a WLAN with connecting to all resources on the network. A user may be required to authenticate to the WLAN and then authenticate to other network resources as well.

121

The two primary authentication models used in WLANs are personal and enterprise models. The differences between them will be discussed later.

Key management is the process of provisioning and replacing encryption keys for confidentiality. Whenever encryption is used, the process of the key establishment must be secure. Additionally, the implementation of keys within the encryption solution must be implemented such that keys are not easily discovered. The weaknesses in the early WEP 802.11 security solution were related to weak key management and implementation.

Encryption provides confidentiality and protection against eavesdropping attacks on WLANs. Modern WLANs use AES encryption or RC4 encryption in some older devices, though RC4 is a deprecated solution in the 802.11-2016 standard.

The process of converting data from its normal state to an unreadable state is known as encryption. The unreadable state is known as ciphertext (or cipher data), and the readable state is plaintext (or plain data). The normal way to encrypt something is to pass the data through an algorithm using a key for variable results. For example, let's say we want to protect the number 108. Here is our algorithm for protecting numeric data:

(original data / crypto key) + (3 x crypto key)

Using this algorithm with a key of 3, we come up with this:

108 / 3 + (3 x 3) = 45

In order to recover the original data, you must know both the algorithm and the key. Needless to say, modern crypto algorithms are much more complex than this, and keys are much longer, but this overview gives you an idea of how things work with data encryption.

The remaining portions of this section detail the specific authentication, key management and encryption solutions used in WLANs and those that are now deprecated and should no longer be used as well as those that are transitioning to be removed from modern implementations.

WPA and WPA2

Wi-Fi Protected Access (WPA) and WPA2 are certifications by the Wi-Fi Alliance that are based on security solutions in the 802.11 standard. WPA is roughly equivalent to Temporal Key Integrity Protocol (TKIP) for key management and RC4 for encryption. WPA is not mentioned in the 802.11 standard, but TKIP and RC4 are addressed, and WPA certifies that they are used as specified in the standard. WPA2 is roughly equivalent to the Counter Mode with Cipher Block Chaining Message Authentication Protocol (CCMP) for key management and AES for encryption. WPA2 is not mentioned in the 802.11 standard, but CCMP and AES are addressed, and WPA2 certifies that they are used as specified in the standard.

> You will not be required to remember what the CCMP acronym stands for in any CWNP certification, including CWS. It is essential for the CWS to know that it is a key management protocol used in 802.11 networks.

WPA – TKIP/RC4

TKIP/RC4 and WPA certification was a transitional solution that was introduced in the 802.11i amendment. It was only intended to bridge the gap to CCMP/AES from the very weak WEP and should not be implemented as a primary solution in modern WLANs. The exception to this guideline is the implementation of WPA for compatibility with irreplaceable hardware or with hardware that cannot be replaced at the time of implementation. In such cases, WPA may be required, but it should be implemented with the strictest of security guidelines:

- **WPA-Personal:** Only implement with very long passphrases, such as 20 characters or more. Change the passphrase at an acceptable interval (e.g., every 30 days, every 90 days).

- **WPA-Enterprise:** Implement with a strong EAP type, such as EAP-TLS, EAP-TTLS or PEAP with proper security precautions.

Ideally, only WPA2 will be implemented, and in most office environments, WPA2 can be safely chosen as the only security solution.

TKIP provides the integrity check procedures in a WPA device, and RC4 provides the encryption. RC4 is a stream cipher, which means that it encrypts one bit at a time in a stream to create the output data (ciphertext). TKIP also defined some key management functions, such as the proper mixing of data with the static key to encrypt the actual data. The key used for actual encryption is a 128-bit key with TKIP/RC4.

According to research reported by Mathy Vanhoef in *A Security Analysis of the WPA-TKIP and TLS Security Protocols*, many networks still used WPA as of April 2016. A horribly significant number still used WEP. Table 5.1 shows this reported information.

	Location	Open Networks	WEP	TKIP	CCMP
2014	New Orleans	340	85	848	1443
2014	Leuven	237	289	3850	5096
2016	Leuven	231	176	4364	6963

Table 5.1: Network Security Methods Reported by Mathy Vanhoef, July 2016

Considering the trends in Table 5.1 and the fact that approximately 3 percent of both New Orleans and Leuven networks used WEP (against the total of all four network types), had the data been available for New Orleans in 2016, it is likely that it would have matched the roughly 1.5 percent shown in Leuven. Given how long it has been since WEP has been known to be a horrible solution for Wi-Fi security, this is still a significant number of very vulnerable networks.

Also, the TKIP (WPA) networks only reduced by 3.5 points from 2014 to 2016 (moving from 40.6 percent to 37.1 percent of total networks among the four types). It appears the transition from WEP to WPA2 is still very much underway. The data for Table 5.1 was collected using random scanning, and one would like to think that specific scanning of business networks would result in much higher percentage usage of CCMP (WPA2).

WPA2 – CCMP/AES

CCMP/AES and WPA2 certification is the primary security solution that should be used in WLANs today. It uses strong key management and encryption solutions. The Advanced Encryption Standard (AES) was chosen as the United States government standard for encryption in 2001, and it replaced the older Digital Encryption Standard (DES). It uses key sizes of 128, 192 or 256 bits. As implemented in 802.11, AES uses 128-bit encryption key lengths.

Most enterprise deployments of WPA2 utilize 802.1X standards-based authentication with the Extensible Authentication Protocol (EAP). EAP is a standard model for authentication and is implemented as an EAP method, of which there are many. 802.1X is port-based authentication and depends on the EAP method for actual authentication. 802.1X defines only the conceptual model on which the authentication operates, which includes two virtual ports for each connection: a controlled port and an uncontrolled port. These ports are technically called port entities.

In an 802.1X implementation, the controlled port is used to transmit data, and the uncontrolled port is only used for authentication. Before authentication is completed, the controlled port is blocked. Therefore, data cannot be transmitted or received until authentication has occurred on the uncontrolled port.

For authentication to occur, 802.1X defines three components: the supplicant, the authenticator, and the authentication server. The supplicant is the component that desires access to the network. The authenticator is the component that provides access to the network and to the authentication server. The authentication server is the component that actually performs the authentication, typically a RADIUS server using 802.11 implementations. Figure 5.3 illustrates the 802.1X authentication components.

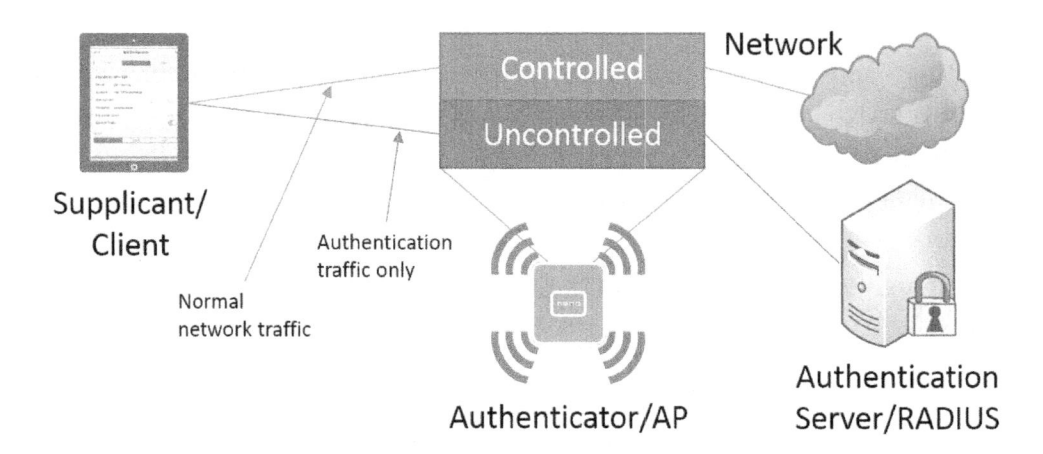

Figure 5.3: 802.1X Components

While 802.1X provides the framework for authentication, it does not provide the actual authentication protocol. This is the work of an EAP method. An EAP method simply defines how the authentication transaction is performed. Figure 5.4 illustrates how an EAP method would fit into the overall 802.1X framework.

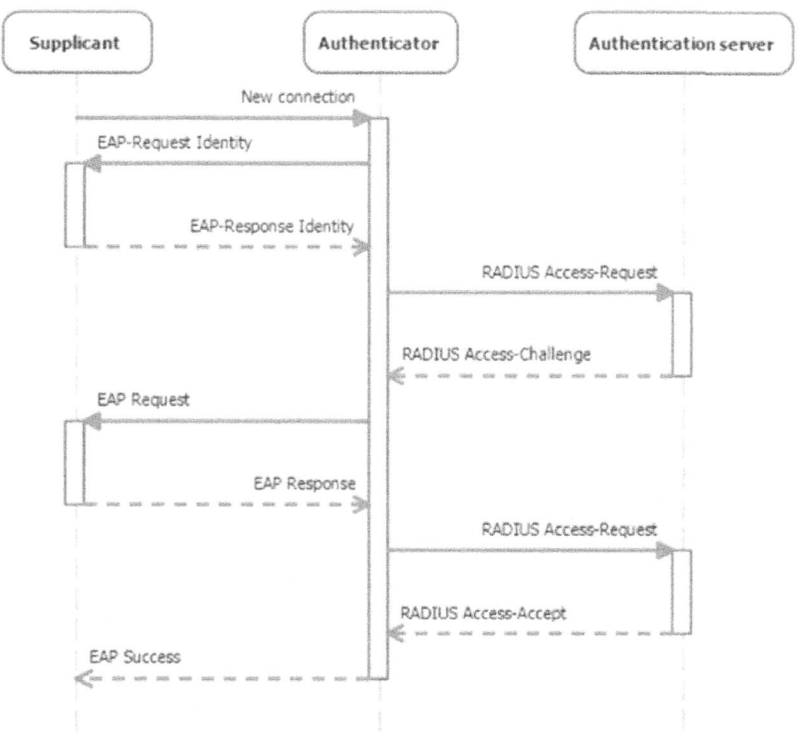

Figure 5.4: EAP Methods in the 802.1X Framework

Knowing the complexities of each EAP method is not required for the CWS exam, but it is important to know that the EAP type used must be supported by both the supplicant (wireless client) and the authentication server (RADIUS). It is also important to know that the materials (data) used to create the encryption keys are generated during the EAP authentication procedure.

An alternative to 802.1X/EAP authentication is pre-shared key (PSK) authentication. PSK authentication uses a passphrase that is converted to the PSK using a known algorithm. The algorithm is defined in the 802.11 standard. This allows you to input a passphrase (e.g., C0MP4NY911) and have it converted to the actual 128-bit hexadecimal key used on the network. PSK is suitable for small businesses or small networks. The PSK must be changed periodically to ensure ongoing security, and with the many clients used in an enterprise network, this management is simply not efficient.

Personal vs. Enterprise

The preceding section referenced small businesses and enterprise deployments. These terms describe the size of the WLAN or the organization in which the WLAN is implemented. The Wi-Fi Alliance has taken the two methods of TKIP/RC4 or CCMP/AES implementation defined in the 802.11 standard and referenced them as Personal or Enterprise. Personal is more suited for small deployments and Enterprise for large deployments, generally speaking. But what is the difference between these two concepts? This section will now explain these concepts.

WPA- and WPA2-Personal use the PSK model for authentication. A passphrase is entered into the AP or controller for the SSID and into the clients. This passphrase is converted to the PSK using an algorithm defined in the 802.11 standard. The PSK is the materials (data) used to generate the encryption keys.

WPA- and WPA2-Enterprise use the 802.1X/EAP model for authentication. The EAP authentication results in the creation of materials called the Master Session Key (MSK). These materials (data) are used to generate the encryption keys. The Enterprise model requires an authentication server, as stated previously, which is typically a RADIUS server. Because of this, it is not practical in many smaller implementations; therefore, the Personal model is often used instead.

> *Remote Authentication Dial-In User Service (RADIUS)* is an IETF standard defined in Request for Comments (RFC) 2865. It was originally created in the days of modem pools for dial-in users; therefore, its name is not as useful today, but its functionality remains very useful.

The general guideline is that smaller networks use WPA2-Personal and larger networks use WPA2-Enterprise. The line of differentiation is not a fixed number of clients but rather one of capability. If the organization is capable of providing for WPA2-Enterprise, it should; if it is not, it should use WPA2-Personal. This capability is defined by technological resources and budgets. Given that large enterprises typically have the resources and the budget, they tend to implement WPA2-Enterprise.

It is important to know that even large enterprises may implement WPA2-Personal (or in worst case scenarios, WPA-Personal) for specific device types. For example, some client devices only support the Personal model. If for some reason, these devices must be used, PSK must be provided on an SSID. The APs that must support the PSK devices will often run two SSIDs: one for Personal and the other for Enterprise.

WPA2-Enterprise introduces the complexity of choosing an EAP method. Several EAP methods are commonly supported in today's WLANs, but the selected method will depend on client compatibility and security needs. Some EAP methods require certificates, which are digital documents that contain a public key and are used to establish tunnels for authentication. When client certificates are required, organizations must implement a Public Key Infrastructure (PKI), which includes setting up certificate authority servers and introduces extra cost and complexity. However, client certificates in addition to server certificates provide authentication that is generally considered more secure. The complexity introduced is that of installing a PKI, installing certificates on servers, and, when required, installing certificates on clients.

> A *tunnel* is a logical entity established between two endpoints that may or may not use encryption. When a secure tunnel is established, encryption is used, and internal authentication (that which transpires in the tunnel) is secured.

The following EAP types are common in WLANs and are considered industry standards:

- EAP-TLS: requires client and server certificates
- EAP-TTLS: requires only server certificates
- EAP-PEAP: requires no certificates but supports them for the inner authentication

- EAP-SIM: works with mobile phones

> An additional security solution that is sometimes used is a Virtual Private Network (VPN). A VPN establishes a secure tunnel between two endpoints. A VPN is often used for security when connecting a laptop to a public hotspot that lacks encryption.

Deprecated Security

According to the 802.11-2016 standard, WEP, WPA (TKIP/RC4) and Shared Key authentication are all deprecated. The standard states the following:

> *Clause 5.1.2:*
>
> *The use of WEP for confidentiality, authentication, or access control is deprecated. The WEP algorithm is unsuitable for the purposes of this standard.*
>
> *The use of TKIP is deprecated. The TKIP algorithm is unsuitable for the purposes of this standard.*
>
> *Clause 12.3.3.1:*
>
> *Shared Key authentication is deprecated and should not be implemented except for backward compatibility with pre-RSNA STAs.*

A deprecated technology should not be planned for current or future WLAN implementations. These security solutions are deprecated because they do not offer sufficient security for modern WLANs. Notice that the three deprecated security options are stated as WEP, TKIP and Shared Key authentication.

Wired Equivalent Privacy (WEP) was included in the first release of the 802.11 standard in 1997. It used the RC4 encryption algorithm but implemented it with improper key mixing. The result was an encryption system that could be cracked in under five minutes by the early 2000s. Because of this, the 802.11i amendment introduced Robust Security Network Associations (RSNAs), which use a four-way handshake for key establishment, stronger encryption key mixing and improved encryption algorithms. WEP should not be used in any WLAN implemented today.

TKIP was introduced in 802.11i-2004 as a transitional security solution for hardware that was unable to run CCMP/AES. It was only intended as a stop-gap until new hardware could be implemented with more processing capabilities. However, as previously illustrated in Table 5.1, many WLANs still use WPA, which uses TKIP/RC4. The reason is often for backward compatibility with client devices that do not support WPA2. However, WPA is often used simply because the implementers do not know better. WPA should not be a planned security solution for any modern WLAN because it has been deprecated in the standard.

Additionally, the 802.11n amendment states that "an HT STA shall not use [...] TKIP to communicate with another HT STA." Therefore, if TKIP/TC4 is used with 802.11n APs and clients, it will force them to act as 802.11a or 802.11g devices, resulting in a maximum data rate of 54 Mbps. The 802.11ac amendment further states, "NOTE – Because a VHT STA is also an HT STA, the elimination of TKIP also applies to VHT STAs." Because of this, both 802.11n and 802.11ac devices that are configured to use WPA will act as an 802.11a or 802.11g device. This result is clearly not desired, and therefore, WPA2 should be used in all modern networks. Figure 5.5 shows these statements as they exist in the 802.11-2016 update of the standard.

> Devices that are designed according to the standard will comply with all standard constraints. However, you may, at times, encounter devices that ignore rules like disallowing TKIP for HT and VHT operations. Such vendors should be avoided when possible.

Shared Key authentication was included in the first release of the 802.11 standard in 1997. It was based on WEP, which is weak in and of itself, but it also introduced a problem in the authentication algorithm. Challenge text was used, and it was sent as clear text across the network. Given that the challenge was sent as clear text and the response was the encrypted text, it made cryptanalysis an easy and simple process. The revelation of the WEP key was made quite simple.

For the basic reasons outlined here, WEP, TKIP and Shared Key authentication should be avoided. WEP and Shared Key authentication should never be used. TKIP (WPA) should only be used in rare cases where it is impossible to use CCMP/AES (WPA2).

> **IEEE Std 802.11-2016**
> **IEEE Standard for Information Technology—Local and Metropolitan Area Networks—Specific Requirements**
> **Part 11: Wireless LAN MAC and PHY Specifications**
>
> of the ESS. One of the possible solutions to this problem might be for the STA to send or receive only frames whose final DA or SA are the AP and for the AP to provide a network layer routing function. However, such solutions are outside the scope of this standard.
>
> An HT STA shall not use either of the pairwise cipher suite selectors: "Use group cipher suite" or TKIP to communicate with another HT STA.
>
> NOTE—Because a VHT STA is also an HT STA, the elimination of TKIP also applies to VHT STAs.

Figure 5.5: TKIP Disallowed for 802.11n and 802.11ac Devices when Using the HT or VHT PHY

BYOD

Bring Your Own Device (BYOD) refers to the allowance of employees and other long-term workers in a facility to bring their devices and connect them to the network. They may use the devices for personal use, but the primary intention is to allow personal devices for use in their work as well. When planning for BYOD, several basic factors must be considered, and control of devices and software must be planned as well. This section introduces these concepts.

BYOD Basics

Today, most users utilize multiple devices at work and at home. At the minimum, it is common for users to have a computer and a mobile phone. In addition, many users have tablets. In fact, in a 2012-Q4 survey (the Forrsights Workforce Employee Survey), 53 percent of users were found to use three or more devices. With each user using multiple devices, we no longer think that a single cable can be sufficient for each desk. Instead, we have to consider how to provide wired (in most cases) and wireless connections in each user's work area.

In addition, usage volume remains an issue. Users reported in the same survey that they used seven or more apps 82 percent of the time for work. Apps, particularly those in the mobile device category, are well known for using network resources. With more apps using resources on multiple devices, we must consider how to handle utilization issues – both in the RF spectrum and in the data throughput capabilities.

Finally, this survey reported that 37 percent of workers said that they work from three or more locations. This reality may demand remote access to the organization's network, remote administration and even remote wipe or disenrollment from the network. This threefold situation (multiple devices, multiple apps, and multiple locations) demands that we consider how to address it. The current and most popular solution is Mobile Device Management. Keep in mind that the data from this survey dates back to 2012: Today users use more devices than ever before.

A BYOD WLAN is no different from a standard office WLAN from the perspectives of RF coverage and capacity. The areas of change are in device on-boarding, device management, and software deployment. If an employee is to use her device on the network for business purposes, the device must have the appropriate software installed. Licensing issues must be addressed, and a software deployment model should be in place. However, this part is beyond the scope of the CWS exam and this book. Here, the focus is on determining which devices should be allowed on the network.

To determine the devices that should be allowed on the WLAN, consider the following issues:

- Older devices may introduce security and performance issues to the network.

- Newer devices may not be fully supported by the BYOD management solution.

- Data leakage must be addressed when allowing organizational data to be placed on personal devices.

The first issue is the security and performance concerns that may be introduced to the network due to older devices. If an older device has not been updated, it may be more susceptible to viruses, worms and other malware (including ransomware attacks that have become popular). This security concern can be addressed by using a Network Access Control (NAC) solution that allows quarantining of devices until they comply with a minimum security baseline. For example, you can place devices that do not comply with the security baseline policy onto a network that allows for device updates but no other network access. Such NAC solutions are available from several vendors and typically involve a moderately complex structure of authentication servers, policy servers, and update servers.

Performance issues are introduced with older devices when they use early radio chipsets that support only 802.11a or 802.11g (or even worse, 802.11b). These devices slow the entire network down because compatibility modes must be used and the older devices take longer to transmit their data. All other devices must wait during these transmissions. No immediate solution for the problem of older PHYs is available. It is simply something that must be accommodated with the implementation of more APs. An alternative to using more APs is simply to disallow any device that does not use an 802.11n radio/chipset or better.

The second issue is supporting the newest operating systems and firmware in the BYOD management solution, like a Mobile Device Management (MDM) package. When users bring the newest devices into the network, the MDM solution may not yet support all features on those devices. Other than disallowing the devices, the only solution is to wait for the MDM vendor to update their software to support the latest devices and firmware or operating systems in those devices.

The final area of concern is data leakage. When allowing personal devices on the network, you must address the issue of organizational data stored on those devices. You can simply choose to disallow it, but that will prevent the users from doing any real business work on their personal devices. A better solution may be to utilize containerization. Many MDM solutions support creating containers on mobile devices that house all organizational data. The containers are encrypted so that data is better protected. Additionally, if the device is lost or stolen, the MDM software can execute a remote wipe command to delete all sensitive data from the device.

In addition to the data stored on mobile devices, another concern is the use of the camera or microphone to obtain data from the organization. If the organization has very sensitive data that is accessible from specified areas, some MDM solutions will allow you to disable the camera and/or microphone on managed devices that enter the area. This configuration allows the devices to be used in the area for other work functions, but it prevents the users from taking pictures of documents or recording conversations that might leak sensitive data. More information about MDM solutions will be covered in the next section: *Device and Software Control Solutions*.

> Remember that MDM solutions can disable features on managed devices. These features include the camera or microphone, installation of software, and more. Such capabilities provide the administrator with greater control of mobile devices in the enterprise.

Device and Software Control Solutions

The primary solution for BYOD implementations is MDM, as referenced in the preceding section. This section introduces MDM solutions with more details about their features and operations.

An MDM solution is a centralized management engine that allows for the onboarding, management and decommissioning of mobile devices. The MDM solution is used to manage personal devices, but the same implementation can manage enterprise devices as well. The MDM managed components include the following:

- **Devices:** laptops, tablets, phones, etc.

- **Applications:** apps from application stores, third-party apps, custom developed apps, etc.

- **Configurations:** typically policy- or profile-based configuration settings for applications and operating systems

- **Functionality:** involves restrictions of features (for example, disabling a camera on a mobile phone)

MDM features include the following:

- Registration
- Pre-registration
- Onboarding
- Profile management
- Device wipes
- Application management

Registration is the process used to grant access to the network for a mobile device and to provision that device for proper use on the network. Registration is also called enrollment and may be performed in different ways, including the following:

- **Pre-registration:** The device is manually registered with the MDM solution before deployment to the user. This method is used for organization-owned devices.

- **Onboarding:** The device is registered with the MDM solution through interactions between the device and the MDM server with possible involvement of user decisions. This method is used for both organization-owned devices and user-owned devices.

Profile management is the process of creating and administering profiles. An MDM profile is a collection of settings that can be applied to specific mobile device types or specific mobile devices. These settings include device restrictions, e-mail settings, VPN settings, LDAP directory service settings and Wi-Fi settings. The administrator must create the profiles and understand how they are assigned to a device for installation within a given MDM solution.

Over-the-air (OTA) management provides the ability to send application updates, OS updates, configuration changes and profile changes without requiring a wired synchronization. For example, it is common to require that a certificate is placed on the device for some EAP types (e.g., EAP-TLS). This certificate may be provisioned

OTA instead of requiring pre-staging. Therefore, OTA management can begin at the registration point and continue through the life of the device.

OTA may be provided through the Wi-Fi network, and it may be provided through push services when not connected to the WLAN. Push services are provided through APIs for IOS, Android and Windows devices. With push services, updates may be transmitted to devices even when they are not connected to the enterprise networks. Additionally, tasks such as device wipes may be performed.

Device wipes are used to delete information on the mobile devices and usually fall into one of two categories:

- **Enterprise Wipe:** Only enterprise settings, files, and apps are removed.
- **Device Wipe:** The entire device is reset to the factory defaults, and all private data is removed.

Application management involves both provisioning and restrictions. Application provisioning is the process used to place applications onto the mobile devices. Application restrictions are used to disable entire applications, features of applications and sometimes features of the mobile OS.

When considering an MDM solution, ensure that it meets the needs of the organization in these areas.

Guest Access

Guest access is often provided so that visitors to the facility can gain Internet access. Guest access is typically provided in one of two ways:

- Separate SSIDs for each network: one for the enterprise WLAN and one for guest access.
- One SSID for both networks with the use of user provisioning and on-boarding for access to network resources.

With the first option, the SSIDs are typically mapped to VLANs that restrict the capabilities of the WLAN clients on the network. For example, the enterprise

WLAN SSID is mapped to a VLAN that provides access to local network resources, while the guest WLAN SSID is mapped to a VLAN that only provides Internet access.

With the second option, the provisioned user account is constrained such that only approved resources can be accessed. A form of captive portal is used when connecting to the WLAN so that users can be authenticated.

> A *captive portal* is a network solution that forces users to visit a web page before they can use the network. This web page provides authentication options, usage policy agreement forms, or simply displays a starting point for network use.

When building guest networks, VLANs are often used to segregate them from the rest of the network. However, a guest network can be open or secured. In an open guest network, the users are not required to authenticate as users but are simply instructed to connect to the guest network (based on an SSID). The guest network is often unencrypted, and it functions much like a traditional hotspot.

In a secured guest network, users are required to know a web-based authentication code or WPA-Personal or WPA2-Personal keys, or they must register as guests. Regardless of the steps, the end result is that they are connected with strong, modern Wi-Fi security.

> A *Virtual Local Area Network (VLAN)* is a technology that operates in switches and allows for traffic separation. For example, a device in VLAN 10 cannot communicate directly with a device in VLAN 20 on the same switch without passing through some routing mechanism.

When implementing a guest WLAN, it is best to separate it from the existing network. This can be accomplished using VLANs with an enterprise network, or physical device separation could be performed using routers. Boundaries can be implemented using firewalls (physical) or VLANs (virtual). You can also use tunnels from the APs or controllers to the demilitarized zone (DMZ) for access to the Internet.

> A *demilitarized zone (DMZ)* is a network boundary that is separated from the internal network and the internet using routers or firewalls. It acts as a boundary network for the placement of Internet servers and a point of exit for tunnels used by guests of the network.

Chapter Summary

In this chapter, you learned about the security solutions that are available in standards-based WLANs. You learned about WPA and WPA2 and that all authentication and key management solutions apart from WPA2 that are commonly available from the 802.11 standard are now deprecated. Additionally, you learned about special cases that require security consideration in WLAN deployments, including BYOD and guest access. In the next chapter, you will learn about enhanced 802.11 functions such as wireless mesh networks, Quality of Service, Dynamic Rate Switching and fast and secure roaming.

Points to Remember

Remember the following important points:

- Common WLAN vulnerabilities include eavesdropping, DoS, management interface exploits, encryption cracking, authentication cracking, MAC spoofing, peer-to-peer attacks and social engineering.

- Confidentiality is the concept of keeping private information.

- RBAC implements authorization using groups or roles.

- Key management is the process of provisioning and replacing encryption keys used for confidentiality.

- WPA is based on TKIP/RC4 and WPA2 is based on CCMP/AES.

- WPA- and WPA2-Enterprise use 802.1X/EAP authentication.

- WPA- and WPA2-Personal use a pre-shared key (PSK), usually based on a passphrase, to authenticate devices and provision encryption keys.

- WEP, Shared Key authentication, and TKIP are all.

- When WPA is used in an 802.11n/ac deployment, it effectively causes the STAs to function as 802.11a or 802.11g STAs, which greatly diminishes performance.

- BYOD is a reference to the users that utilize their personal devices for business work on the organization's WLAN.

- MDM can be used to manage BYOD and enterprise devices.
- Guest networks may be separated from enterprise networks with VLANs or tunnels.
- SSIDs can be mapped to different VLANs to constrain the connecting device's network access.

Review Questions

1. What security solution is used to protect against eavesdropping attacks?
    a. Passwords
    b. Accounting
    c. Encryption
    d. Client firewalls

2. What authorization technique uses groups or roles to control actions of WLAN clients?
    a. RBAC
    b. VPN
    c. WPA
    d. WPA2

3. What encryption solution is used with WPA?
    a. AES
    b. RC4
    c. DES
    d. Triple-DES

4. What encryption solution is used with WPA2?
    a. AES
    b. RC4
    c. DES
    d. Triple-DES

5. TKIP/RC4 is the foundation for what Wi-Fi Alliance certification?
    a. WPA
    b. WPA2
    c. Voice Enterprise
    d. Voice Personal

6. Which one of the following is a deprecated security solution in the 802.11 WLAN standard?
    a. AES
    b. CCMP
    c. WEP
    d. IPSec

7. What is used to manage mobile devices in BYOD implementations?
    a. AES
    b. MDM
    c. RC4
    d. DRS

8. What WLAN vulnerability takes advantage of human manipulation to gain access to the network?
    a. Authentication cracking
    b. Encryption cracking
    c. Social engineering
    d. Management interface exploits

9. What is used by WPA2-Enterprise for authentication and key provisioning?
    a. Pre-shared key
    b. RBAC
    c. MDM
    d. 802.1X/EAP

10. What string is usually passed through a defined algorithm to generate the PSK used in WPA2-Personal deployments?
    a. Passphrase
    b. SSID
    c. RSSI
    d. Channel

Review Answers

1. **C is correct.** Encryption provides confidentiality and is used to protect against eavesdropping.

2. **A is correct.** Role-Based Access Control (RBAC) allows for restrictions of WLAN clients based on group membership or roles.

3. **B is correct.** WPA is roughly equivalent to TKIP/RC4, so the encryption used is RC4.

4. **A is correct.** WPA is roughly equivalent to CCMP/AES, so the encryption used is AES.

5. **A is correct.** WPA certifies that devices implement TKIP and RC4 based on the 802.11 standard.

6. **C is correct.** WEP, Shared Key authentication, and WPA are all deprecated in the 802.11-2016 standard.

7. **B is correct.** Mobile Device Management (MDM) solutions provide for the management of BYOD and enterprise mobile devices.

8. **C is correct.** Social engineering uses manipulation tactics to obtain information that should not otherwise be provided.

9. **D is correct.** WPA- and WPA2-Enterprise use 802.1X/EAP for authentication and key provisioning. The source materials from the 802.1X/EAP authentication are used in the four-way handshake to generate the encryption keys.

10. **A is correct.** A passphrase is passed through a defined algorithm to generate the pre-shared key (PSK) in most implementations.

Chapter 6 – Enhanced 802.11 Functions

Objectives Covered

2.6 Describe features of enhanced 802.11 functions
- 2.6.1   Mesh
- 2.6.2   Quality of Services (QoS)
- 2.6.3   SISO vs. MIMO
- 2.6.4   Dynamic Rate Switching (DRS)
- 2.6.5   Backwards compatibility

3.5 Explain the requirements of fast and secure roaming for non-technical professionals
- 3.5.1   Latency requirements for streaming communications
- 3.5.2   Pre-authentication
- 3.5.3   Key caching methods

The basic features of 802.11 were covered in preceding chapters. Also, some of the features explained in this chapter have been referenced in earlier chapters. However, the focus of this chapter is to explain in greater detail features like mesh, SISO vs. MIMO, Quality of Service, Dynamic Rate Switching, and Fast Secure Roaming. You will also learn more about backward compatibility in 802.11 networks.

Enhanced Features of 802.11

Several enhanced features are provided in the 802.11 standard. Understanding their use and capabilities is beneficial to the CWS. These features include mesh, SISO and MIMO communications, Quality of Service, Dynamic Rate Switching, and backward compatibility.

Mesh BSS

Most 802.11 networks are implemented in a type of client/server model where the mobile device is the client, and the AP is the server. However, other types exist including ad-hoc networks and mesh BSS networks. An ad-hoc network, which is known as an independent BSS (IBSS) in the standard, is a network that is created directly between mobile devices. The mobile devices communicate with each other without the use of an AP. This network type will not be explored in more depth here, but as a CWS, you should know that an IBSS allows for direct communications between mobile devices without the use of an AP.

The mesh BSS (MBSS) is a BSS with interconnecting APs that allow for client connections as well. The following is a definition of an MBSS from the 802.11 standard:

> *A BSS that forms a self-contained network of mesh STAs that use the same mesh profile. An MBSS contains zero or more mesh gates and can be formed from mesh STAs that are not in direct communication.*

From this definition, the reason for mentioning the IBSS network becomes clearer. An IBSS network only includes stations (STAs) that can communicate directly with each other, while an MBSS does not have this requirement. If STA A can communicate with STA B, and STA B can communicate with STA C, even though STA C cannot communicate with STA A directly, they can all still form an MBSS.

Additionally, while an IBSS will not connect to a distribution system (DS), such as an Ethernet network, an MBSS may connect to a DS. An MBSS can use mesh gates to connect outside of the mesh to another network (e.g., an Ethernet network) as the DS. Interestingly, an MBSS can act as the DS for other BSSs to provide access to STAs. All mesh nodes appear to share the same MAC layer even if they are not in range of each other, and mesh nodes can be sources, sinks or propagators for traffic.

> A *distribution system (DS)* is a system that is used to provide connections among WLAN BSSs and integrate with a local area network (LAN), such as an Ethernet network. An ESS is formed when BSSs are interconnected across a shared DS.

A mesh source is the originator of the traffic into the mesh structure. The sink is the destination of the traffic, and the propagators simply forward traffic from the source to the sink.

Figure 6.1 from the 802.11 standard illustrates the concepts in a network that employ MBSS structures. You can see two MBSSs named mesh BSS 1 and mesh BSS 2. Notice that the mesh gates connect the MBSSs to other networks. Also, note that the AP that provides the infrastructure BSS 25 is the same device that acts as a mesh STA and mesh gate into mesh BSS 2. The point of this illustration is to show that a single AP can act as a service provider for a traditional BSS while also running the MBSS as a mesh station or mesh gate.

Several vendors provide mesh functionality in their APs. Some implement a proprietary solution, and others use the MBSS model as described in the 802.11 standard. In most cases, one radio is used for the service of 802.11 clients, and the other is used to participate in the MBSS. With the modern advent of tri-band radios and as they improve in functionality, APs may support two radios for client access and one for MBSS participation or two for MBSS participation and one for client access.

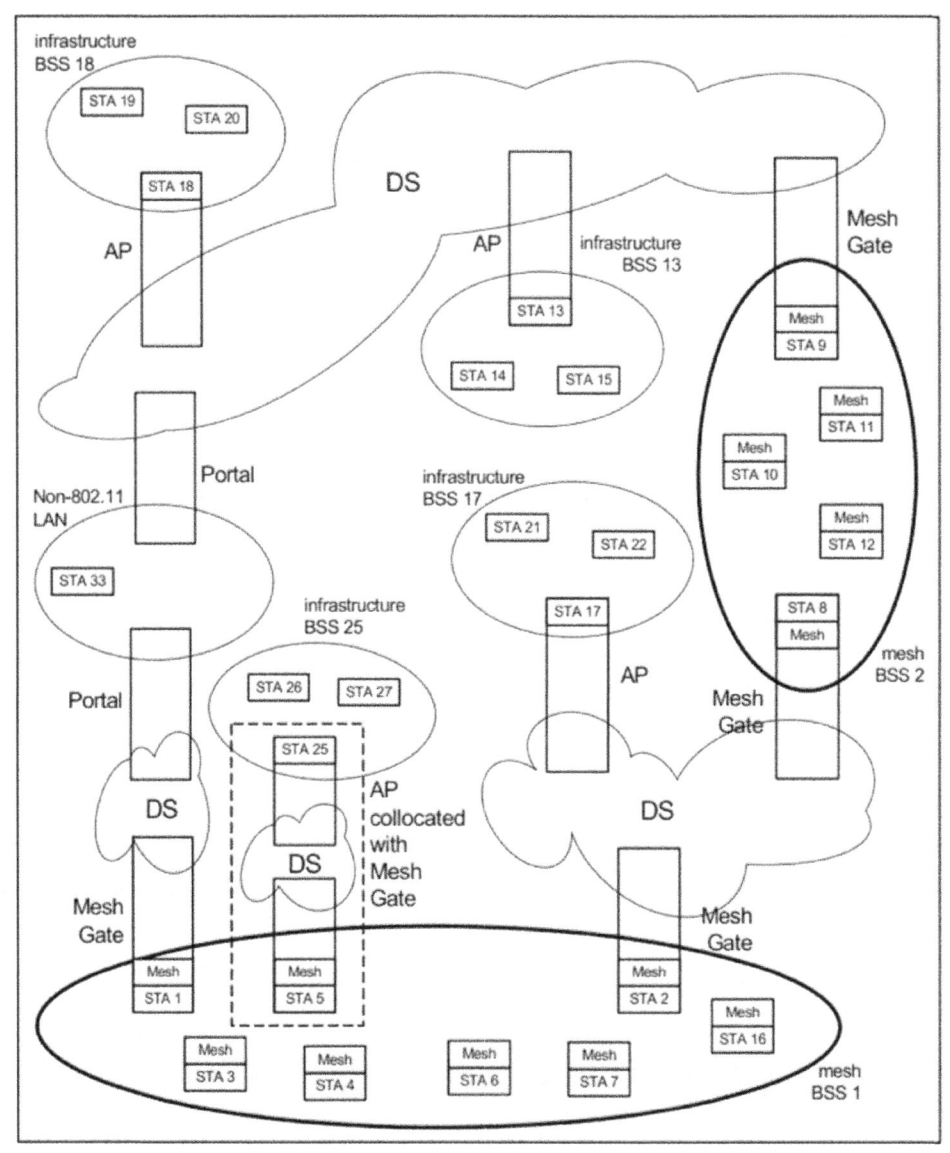

Figure 6.1: MBSS Illustrated in the 802.11 Standard

> The 802.11 standard introduced mesh capabilities in the 802.11s amendment. This is incorporated into 802.11-2016 and is part of the current standard. It defines the elements that make up the mesh profile contained in beacons and probe request/responses.

For the CWS candidate, it is most important to remember that an MBSS differs from a BSS in that it provides interconnections among the STAs such that they all appear to be on the same MAC layer, even if they are not in range of each other. An MBSS supports all standard 802.11 features like multiple spatial streams, and all supported PHYs.

SISO vs. MIMO

802.11 devices can use two kinds of fundamental communications: SISO and MIMO. SISO communications are single-input and single-output and use one antenna for transmission and reception. MIMO communications are multiple-input and multiple-output and use two or more antennas for transmission and reception. This section explores the functionality of both communication methods as well as their issues.

The earliest 802.11 networks supported SISO communications, and the newest networks do as well. Figure 6.2 illustrates the concept of a SISO transmitter. A single antenna transmits to a single receiving antenna. 802.11n and 802.11ac devices still support SISO communications. These communications are also called single stream communications. 802.11 (DSSS), 802.11b (HR/DSSS), 802.11a (OFDM) and 802.11g (ERP) were all strictly SISO systems. These often used antenna diversity, but they did not support multiple spatial streams or MIMO.

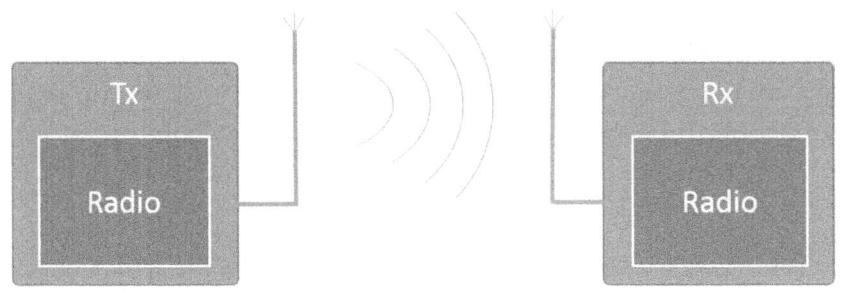

Figure 6.2: SISO Transmissions

Antenna diversity is where one antenna is used to transmit and receive actual data frames, but the best antenna is chosen for each communication, and the decision is made at the time of reception. The antenna that sees the best signal during early communication is used to receive the rest of the frame. When transmitting back to the STA from which it received the frame, the same antenna that received the best signal is used, assuming that it will also send the best signal to the STA.

A significant problem for SISO systems is multipath effects. As you learned in previous chapters, multipath effects occur when the RF signal reflects off of objects in the area and results in multiple copies of the signal arriving at the receiver. When these signals are significantly out of phase with each other, the received signal is degraded or canceled, which causes poor performance or the inability to receive the data. The impact may range from lower data rates up to a complete loss of connection in extreme cases.

Typically, indoor networks, especially those in warehouses and manufacturing environments, are prone to high multipath effects. This is a problem in SISO systems, and the traditional solution to this issue was to implement antenna diversity. Two antennas, separated by half a wavelength, were installed in the AP and/or client (mostly in APs though). This small separation between the antennas meant that one antenna would receive a better signal than the other. The antenna with the best signal quality would be used for communications.

When one device in the link supports MIMO and the other supports only SISO, the connection will downgrade to the lowest common denominator. The result will be a SISO transmission and the lack of multiple spatial streams regardless of the support for it in one of the devices.

Interestingly, MIMO transmissions take advantage of multipath effects. The multiple paths allow the transmitter to send multiple streams to the receiver. To accomplish this, the transmitter and receiver must support some common number of radio chains. If one end of the link supports four radio chains and the other supports only two, the link will be downgraded to the lowest common denominator (i.e., two radio chains and therefore two spatial streams).

A *radio chain* is a path from the radio to an antenna. Devices that support multiple radio chains have multiple antennas. Each radio chain can send and receive a stream of data, and this allows for increased data rates.

If one spatial stream can accomplish 72 Mbps, then two can accomplish 144 Mbps, which effectively doubles the data rate. Today, most client devices support from 1 to 3 spatial streams, and the majority of client devices support 1 or 2 spatial streams. Remember that each antenna requires power to radiate the signal. Therefore, more spatial streams require more power. Given that mobile devices run on batteries, using fewer spatial streams balances data speed with battery life. For this reason, many mobile clients support only 1 spatial stream, and some support 2, but very few support 3, with the exception of some laptops.

Figure 6.2 illustrates MIMO communications. In the illustration, both the transmitter and the receiver support 2 spatial streams (radio chains) and can effectively double the supported data rate given sufficient signal strength.

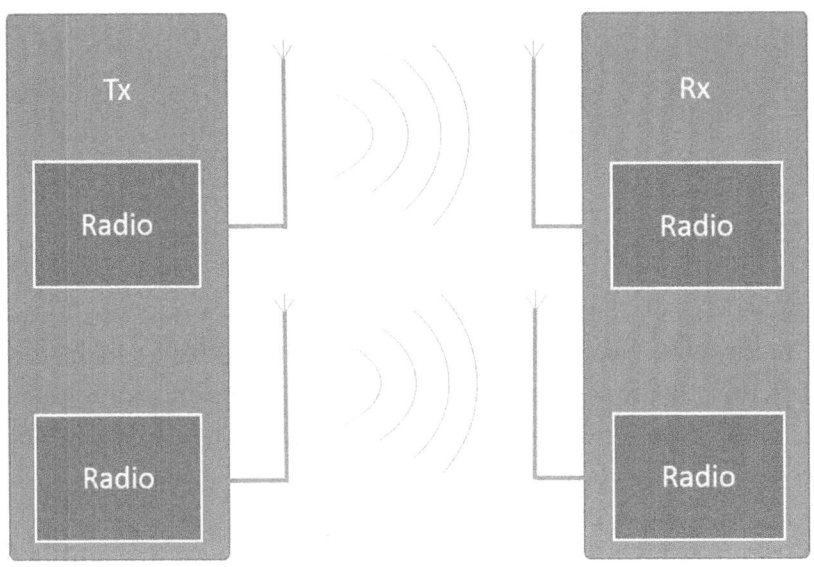

Figure 6.3: MIMO Transmissions (2 streams)

When documenting the number of supported spatial streams, a common nomenclature is used. Three values indicate the transmit radio chains, the receive radio chains, and the supported spatial streams in that order. For example, if a device is rated at 3x3:3, it indicates that it supports 3 transmit (Tx) radio chains, 3 receive (Rx) radio chains, and 3 spatial streams. However, a device rated at 3x3:2 supports 3 Tx and Rx radio chains but only 2 spatial streams. In such an implementation, the third radio chain may be used for antenna diversity.

> For the exam, remember that the nomenclature for radio chains and spatial streams is *Tx radio chains* x *Rx radio chains* : *spatial streams*. As an example, a device supporting two Tx and Rx radio chains and one spatial stream would be defined as 2x2:1.

Quality of Service (QoS)

QoS is implemented in wired and wireless networks to give priority to packets that have a low tolerance for delay, such as VoIP packets. In wired networks, the QoS is closer to guaranteed priority, while in wireless networks, it is a probabilistic priority. This statement simply means that wireless networks allow for the prioritization of packets at the transmitter, but the transmitter must still wait its turn for access to the medium (RF). A wireless STA cannot transmit while another transmission is seen on the medium.

802.11 networks implement QoS using prioritized wait times. Packets that have a higher priority have a lower initial wait time before they can transmit. Packets with lower priority have longer wait times. The end result is that packets with a higher priority get to the point of transmission faster and then must only wait for the medium to clear.

This QoS happens within each STA since each STA has internal timers it must use before it attempts a transmission. The STAs with higher priority packets will count these timers down faster than STAs with lower priority packets because the countdown timers start at lower numbers for higher priority packets. Figure 6.4 illustrates the QoS operation within a STA.

In Figure 6.4, you can see that voice packets would receive a timer in the range of 0-3, while data packets would receive a timer in the range of 0-15. When packets fail to be received in the wireless medium, this range can increase, but for the CWS, it is sufficient to know these starting ranges. Because the voice packets must count down starting from a maximum of three and data packets must count down starting from a maximum of 15, it is more likely that voice packets will expire the countdown timer faster and be the next packet in the queue for transmission.

Given that voice packets have a low tolerance for the delay on the network, it is important to ensure that QoS is enabled on the APs and clients involved in voice communications. Fast roaming is also important for voice packet transmissions and is discussed later in this chapter.

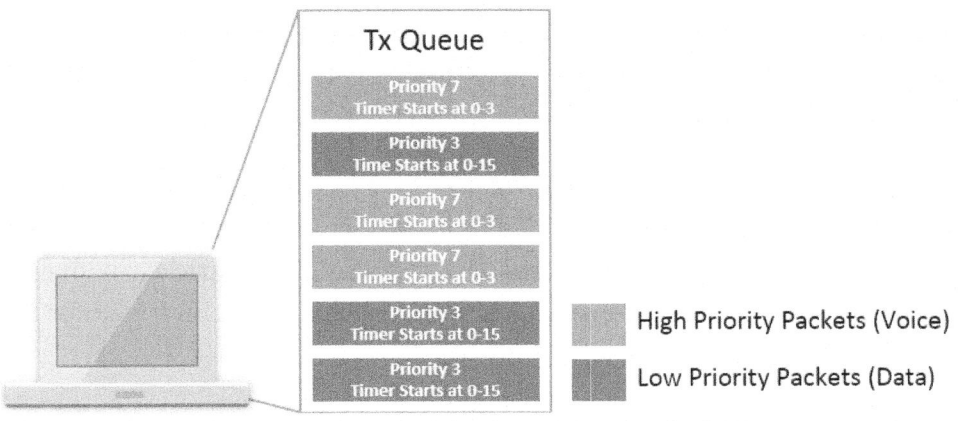

Figure 6.4: QoS Operations in a Single STA

Dynamic Rate Switching (DRS)

DRS is used to accommodate a reduction or increase in signal quality as mobile devices move around within a cell. To achieve a particular data rate, sufficient signal strength is required. To be more specific, sufficient SNR is required, which, as you will remember, is the space between the noise floor and the signal itself. When a sufficient SNR is available, a given data rate can be achieved. As this SNR is reduced, the data rate must be reduced as well. As the SNR is increased, the data rate can be increased as well. This behavior is defined as Dynamic Rate Switching in the 802.11 standard.

> A *cell* is the coverage area provided by an AP. Also called the *Basic Service Area (BSA)* in the 802.11 standard, it is the area in which client STAs can connect with the AP. Within the cell, data rates vary based on the signal quality of the exchanges between the AP and clients.

Figure 6.5 illustrates DRS. The image is based on a single stream client using an 802.11n (HT) radio. The maximum data rate is 150 Mbps. It steps down in the following sequence: 150, 135, 120, 90, 60, 45, 30 and 15. DRS will switch down to these lower data rates as the signal strength continues to decrease, and as the signal strength increases, it will switch back up to higher data rates in the same way. It is also common to skip data rates when switching because of drastic changes in signal quality. For example, a STA may go from 150 Mbps to 90 Mbps instead of switching to the intermediate data rates of 135 or 120 Mbps.

Dynamic Rate Switching (DRS) for 802.11n (HT) with 40 MHz Channel and 1 Spatial Stream

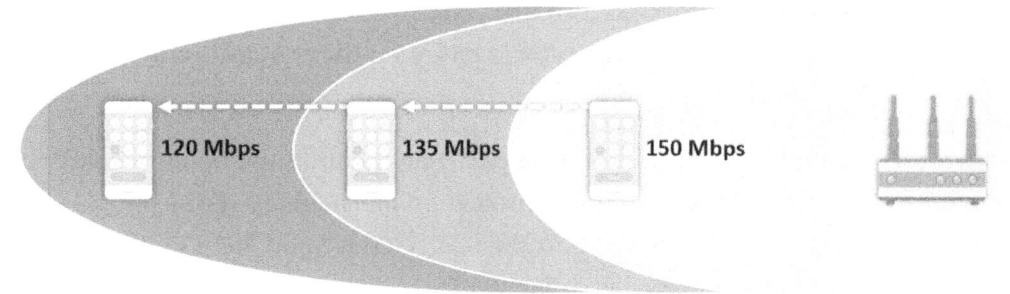

Figure 6.5: DRS Illustrated

It is important to know that one STA at a given location with the same specifications (e.g., number of spatial streams, PHY support) may select a different data rate than another STA at the same location. This behavior is due to varying sensitivity levels on the devices and varying quality of hardware used. Even the exact same device may use different data rates because the radio chains are not necessarily identical, even though they are either iPhone 7s or Samsung tablets of the same model. This is true for all vendors due to the extreme expense involved in accurate calibration of radio chains.

Backward Compatibility

The final concept in this section is backward compatibility. Backward compatibility allows newer 802.11 devices to communicate with older 802.11 devices within some constraints. The constraints on backward compatibility are as follows:

- Devices can only be backward compatible with PHYs that operate in the same band. For example, 802.11n is backward compatible with 802.11g in the 2.4 GHz band. However, 802.11ac only operates in the 5 GHz band, so it is not backward compatible with 802.11g, but it is backward compatible with 802.11a.

- Devices can only be backward compatible with PHYs using similar channel and modulation structures. For example, no modern 2.4 GHz device is backward compatible with Frequency Hopping Spread Spectrum (FHSS) devices, which was a PHY available in the original 802.11 standard. Thankfully, few such 802.11 FHSS devices exist in production today.

- Backward compatibility must be enabled in the APs. It is possible to implement an AP that only supports a modern PHY and does not support older PHYs and therefore older devices. This behavior is based on the configuration and not on the 802.11 standard itself.

Backward compatibility is provided by supporting the modulation of the older PHYs in the current radio and by using protection mechanisms. For example, an 802.11ac (VHT) device will also support the 802.11n (HT) and 802.11a (OFDM) PHYs. The multi-PHY support allows the device to communicate with other 802.11ac devices as well as 802.11n and 802.11a devices.

When older PHYs are present in the cell or detected in surrounding cells on the same channel, the cell must use protection mechanisms. In most cases, the protection mechanisms include one of two solutions:

- **Request to Send (RTS)/Clear to Send (CTS):** With RTS/CTS, the STA wishing to transmit with a newer PHY modulation will first send an RTS frame with an older PHY modulation and receive a CTS frame with the older PHY modulation. The key to the process is that the RTS/CTS frames set timers in all STAs indicating the length of time the medium will be busy. This announcement allows for the transmission of data using the newer PHY modulation that the older devices would not understand.

- **CTS-to-Self:** The CTS-to-Self is used by APs that need to transmit data at a new PHY modulation rate. Rather than wasting time sending an RTS and a CTS, the AP will simply send a CTS, which all connected STAs should be able to process. The CTS frame is sent with the older PHY modulation rate. Now, the medium is clear for the AP to transmit at the higher PHY modulation rate.

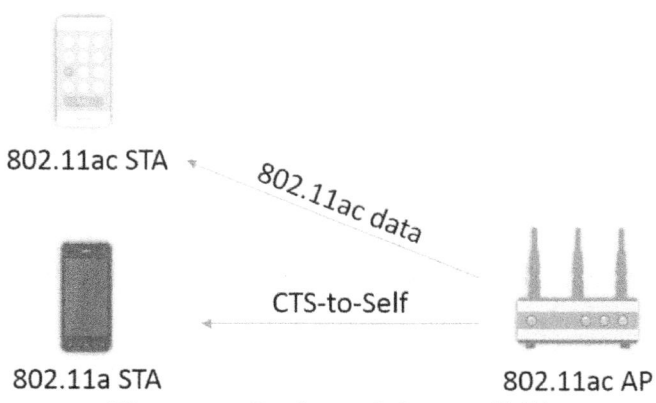

Figure 6.6: Backward Compatibility

For clarity, the goal of the protection mechanism is to prevent older STAs from communicating a modulated signal they do not recognize during the transmission. The RTS/CTS or CTS-to-Self frame sets the timers in the older STAs so that they know they must be silent. Figure 6.6 illustrates the backward compatibility operation. While the CTS-to-Self is received by all STAs in the cell, it is targeted at the older STAs that do not support the newer PHY modulation that is about to be used. The illustration indicates this with the arrow for the CTS-to-Self pointed at the 802.11a STA and the 802.11ac data pointed at the 802.11ac STA. However, you should understand that all frames are seen by all STAs in the cell when transmitted from the AP, even though they may not be able to process them all. The protection mechanism allows older STAs to ignore the bits they do not understand in new modulation transmissions.

Fast Secure Roaming (FSR)

Roaming is the process of moving a client STA from one AP to another. Intra-ESS roaming indicates that the STA is staying in the same ESS, which is usually defined by the matching SSID and shared distribution system. Roaming can occur at two levels in the network layers: the switching layer (called Layer 2 or the Data Link Layer in the OSI model) and the routing layer (called Layer 3 or the Network Layer in the OSI Model). When roaming occurs at the routing layer, the client STA may lose its IP address unless some mechanism is in place to allow it to keep the address, such as tunneling back to the original AP or controller. When roaming occurs at the switching layer, no loss of IP address will occur.

If the BSS is open, meaning it does not use PSK or 802.1X/EAP authentication, roaming will happen very quickly at the switching layer. When PSK (remember, WPA- or WPA2-Personal) is used, roaming still happens very quickly because the PSK is the source material used to quickly perform the four-way handshake and generate encryption keys so that communications can resume. With Open System authentication (i.e., no PSK or 802.1X/EAP), the roaming will usually take around 30 milliseconds (ms). With PSK, the roaming will usually take less than 50 ms. When 802.1X/EAP is introduced, extra communications must occur during roaming with normal operations that often cause the roaming time to exceed 150 to 200 ms.

> The *Open System Interconnect (OSI)* model is a conceptual model for network communication. It uses 7 layers to explain the communication. No protocol directly implements the OSI model, but it is the standard for discussing networking.

Why is all of this discussion of roaming times important? It is important because applications that are intolerant of long delays can cease to function or function poorly when a slow roam (in excess of 50-60 ms) takes place. For example, VoIP typically requires a delay of 150 ms or less in packet transmission from source to destination. If the roaming itself takes more than this 150 ms or close to it, it can result in poor call quality or even dropped calls. For this reason, shortening roaming times is essential.

When open or PSK networks are used, roaming times should not introduce a problem. When 802.1X/EAP networks are used, which is common in enterprise deployments, roaming times must be addressed. It is for this reason that vendors developed a solution called Opportunistic Key Caching (OKC), and the 802.11 working group introduced Fast BSS Transition (FT). Both methods involve caching keys from an initial 802.1X/EAP authentication so that roaming is close to the speeds of PSK roaming. However, one is standard (FT), and the other is proprietary (OKC).

OKC is a sort of standard in that most WLAN infrastructure vendors (those who make APs) support it. However, it must also be supported by the clients. Not all clients support it, so you must be sure to verify its support by your clients. Thankfully, most dedicated VoIP WLAN handsets do support OKC, and it can be used for effective FSR.

FT is the standard method for FSR; however, it was developed later than OKC and is not as widely supported as OKC. As support increases for FT, the transition to standards-based solutions should be made. Many WLAN infrastructure vendors allow supporting both OKC and FT on the same SSID at the same time. This allows supporting a greater number of client devices. It is important to know that some older devices may not be able to connect to an SSID with FT enabled. This limitation is because FT-enabled SSIDs broadcast beacon frames with new information in them. This new information may cause the processing algorithms in legacy devices to break. Be sure to test older devices when enabling FT on infrastructure. If you have incompatible devices, you may be able to use a dedicated SSID for non-roaming-sensitive clients and another SSID for roaming-sensitive clients.

Remember that FT (and not OKC) is included in the 802.11 standard. More and more client devices are beginning to support FT as well as infrastructure devices. It should eventually take over as the primary solution for FSR.

Both OKC and FT are key caching methods. They both cache keys and transfer them among the different APs or controllers so that an entire 802.1X/EAP authentication process is not required. Given that the 802.1X/EAP authentication itself (excluding Open System authentication, which is always required in a roaming process, and the four-way handshake) takes between 50 and 150 ms, removing this requirement greatly decreases roaming delays.

An additional roaming solution, which only reduces roam times by a few ms, is called pre-authentication. *Pre-authentication* is used by client STAs before a roaming event occurs. The client detects other APs in the area to which it could roam and authenticates with them (the first part of Open System authentication) in advance. This action will only save 1 to 5 ms, but it can help if the difference between roaming with good voice quality and roaming without it is a factor of a few ms in a particular implementation.

> Remember that pre-authentication is performed by the client and only saves a few ms. It will not resolve roaming problems with delays in excess of 155 ms in most cases. Other FSR solutions should be sought in such scenarios.

Chapter Summary

In this chapter, you learned about several enhanced features of 802.11 networks. These features included mesh networks, QoS options, SISO vs. MIMO devices, DRS, backward compatibility and fast secure roaming. In the next chapter, you will learn about the detailed features and specifications of APs that are available for modern deployments.

Points to Remember

Remember the following important points:

- An MBSS is a mesh BSS and does not include a standard infrastructure BSS, but it may connect to one through a mesh gate.

- An MBSS is different from an IBSS (ad-hoc) because mesh STAs may be able to communicate with other mesh STAs to which they are not directly connected.

- SISO transmission systems use one antenna to send one spatial stream.

- MIMO transmission systems use multiple antennas to send multiple spatial streams.

- The nomenclature for MIMO radio chains is Tx radio chain x Rx radio chain : spatial streams. An example is that a 3x3:2 device supports 3 Tx and Rx radio chains but only 2 spatial streams.

- When two 802.11 devices communicate, the lowest common denominator related to spatial streams, channel widths and supported PHYs will determine the communications method.

- QoS is implemented in 802.11 networks within each STA and gives probabilistic priority to important packets, such as voice and video communications.

- DRS decreases or increases the data rate in a wireless link based on the signal quality of the link.

- Backward compatibility is achieved by supporting older PHYs that operate in the same band as the device and implementing RTS/CTS or CTS-to-Self.

- OKC is a proprietary standard for FSR, and FT is the 802.11 standard method of FSR.

Review Questions

1. What is a unique defining characteristic of an MBSS?
    a. It supports 802.11ac
    b. It allows communications with nodes that are not directly connected
    c. It is not an infrastructure BSS
    d. It supports multiple spatial streams

2. Through what must an MBSS communicate to access an infrastructure BSS?
    a. Mesh gate
    b. Wired switch
    c. Wired router
    d. None of these

3. What is a 1x1:1 radio system also known as?
    a. MIMO
    b. SISO
    c. Slow
    d. Fast

4. How many receive radio chains does a 4x3:3 device have?
    a. 4
    b. 3
    c. 1
    d. None

5. What is a defining characteristic of a MIMO 802.11 device?
    a. Support for the ERP PHY
    b. Support for the OFDM PHY
    c. Support for a single spatial stream
    d. Support for multiple spatial streams

6. What RF phenomenon does MIMO take advantage of to increase the data rate of the transmission?
    a. Absorption
    b. Multipath
    c. Attenuation
    d. Amplification

7. QoS is guaranteed in 802.11 networks?
    a. True
    b. False

8. What kind of communication needs QoS in wireless networks?
   a. Voice
   b. E-mail
   c. File transfer
   d. Database access

9. What functionality of the 802.11 standard decreases or increases the data rate in a Wi-Fi link based on signal quality?
   a. FSR
   b. OKC
   c. DRS
   d. DFS

10. What kind of frame may be used to implement backward compatibility in 802.11 networks?
    a. Data
    b. Association Request
    c. CTS-to-Self
    d. Acknowledgement

Review Answers

1. **B is correct.** An MBSS allows mesh nodes to communicate with other mesh nodes without being directly connected with them. IBSS network devices can only communicate directly with other IBSS devices. BSS devices can only communicate with the AP directly in most networks and in some networks with other STAs in the same BSS.

2. **A is correct.** A mesh gate is used to allow the MBSS to communicate with an infrastructure BSS. The communication may occur through a device that is both an AP to the BSS and a mesh gate and mesh node to the MBSS.

3. **B is correct.** A device with 1 Tx radio chain, 1 Rx radio chain and 1 spatial stream is a SISO device.

4. **B is correct.** The second part of the nomenclature defines the Rx radio chains; therefore, a 4x3:3 device supports 3 Rx radio chains.

5. **D is correct.** Unlike SISO devices, MIMO devices support multiple spatial streams.

6. **B is correct.** Multipath effects are detrimental to SISO communications and provide the advantage in MIMO communications to allow for multiple spatial streams.

7. **B is correct.** QoS is not guaranteed in 802.11 networks. It is a probabilistic QoS. You could argue that it is guaranteed within the device, but gaining access to the medium is still contention based.

8. **A is correct.** Voice communications require QoS in wired and wireless networks. E-mail, file transfer (FTP for example) and database access do not require QoS in most cases.

9. **C is correct.** Dynamic Rate Switching (DRS) decreases or increases the data rate in a link as the signal quality changes. It decreases or increases to the next available modulation and coding rate as specified in the standard.

10. **C is correct.** Backward compatibility can be performed with RTS/CTS frames or with a single CTS-to-Self frame.

Chapter 7 – Wireless Access Points

Objectives Covered

3.1 Identify AP features and capabilities
- 3.1.1   PHY support
- 3.1.2   Single-band vs. dual-band
- 3.1.3   Output power control
- 3.1.4   Operational modes
- 3.1.5   Multiple-SSID support
- 3.1.6   Guest access
- 3.1.7   Security features
- 3.1.8   Management interfaces
- 3.1.9   Internal and external antennas
- 3.1.10   PoE support

3.2 Describe AP management systems
- 3.2.1   Autonomous
- 3.2.2   Controller
- 3.2.3   Cloud
- 3.2.4   Management systems

Access Points (APs) form the foundation of WLANs in enterprise deployments. They implement BSSs and allow for client STA connectivity. When selecting the appropriate APs, it is important to understand their features and capabilities. In this chapter, you will learn to read an AP specification sheet (spec sheet) and understand the meanings of the various sections. Understanding this information will allow you to select the appropriate AP for a given application. Finally, you will explore various AP management solutions, including autonomous, controller-based, cloud-based and other management systems.

AP Features and Capabilities

APs come in many shapes and sizes. They are designed for internal or external use. They offer internal and/or external antennas. They can be powered by wall outlets or using Power over Ethernet (PoE). This section explains the many features and capabilities of APs.

Figure 7.1 shows several different APs from different vendors. Some are designed for internal use and others for external use. Additionally, some are shown with external antennas; others have internal antennas and do not show visible antennas.

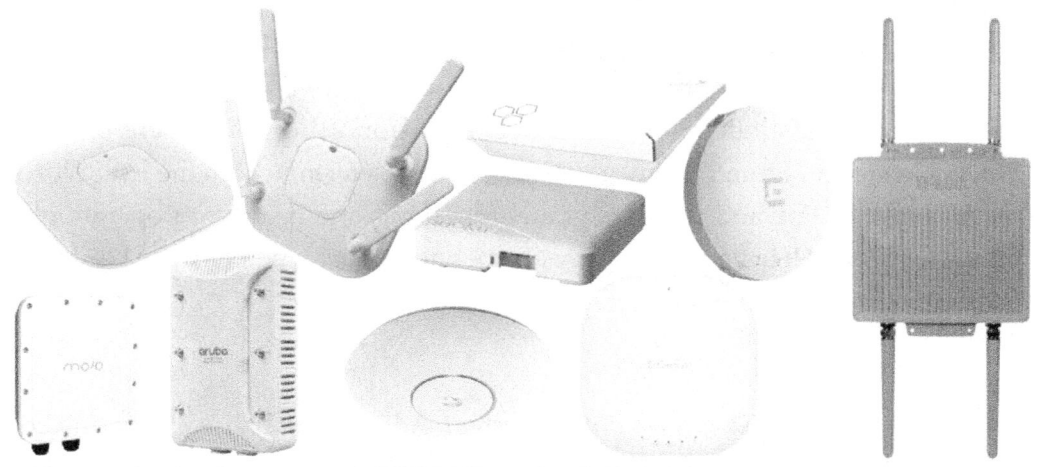

Figure 7.1: Various Aps

As you can see from Figure 7.1, APs vary greatly in appearance, but they also vary significantly with respect to their features. These feature differences are important to understand, even within a particular vendor's line of APs. The rest of this section will explore the various features and capabilities of APs. The spec sheet chosen for the EnGenius AP is not intended as an endorsement for a particular vendor. You should explore the spec sheets for your vendor's APs to better understand their offerings.

A spec sheet provides the specific (as its name implies) details of the features and capabilities of a device. For APs, this will include important characteristics like PHYs supported, bands supported, power options, management options and more. In this section, the spec sheet for the EnGenius EWS370AP indoor AP will be discussed.

> In many cases, the spec sheet is part of the data sheet for an AP. The data sheet often includes additional marketing literature beyond the simple specifications of the device. This information can also be of value in the process of selecting an AP.

Figure 7.2 shows the specifications for the EnGenius EWS370AP from an enclosure perspective. Notice the indication of Ethernet port LEDs, Power LED, Wi-Fi band LEDs (for both 2.4 GHz and 5 GHz), Kensington Security Lock (used to prevent theft of the AP), mounting holes, Ethernet ports supporting PoE, Reset Button and Power Connector. From this image alone, you can determine that the EWS370AP model supports PoE and two gigabit Ethernet ports, and it supports operations in the 2.4 GHz or 5 GHz bands. This image does not reveal whether the AP is dual-band concurrent, but it does support both bands.

> The phrase *dual-band concurrent* indicates that an AP supports both the 2.4 GHz and 5 GHz bands and that it can run both bands simultaneously. Some APs are dual-band but not dual-band

concurrent, though non-concurrent operations is rare in enterprise APs.

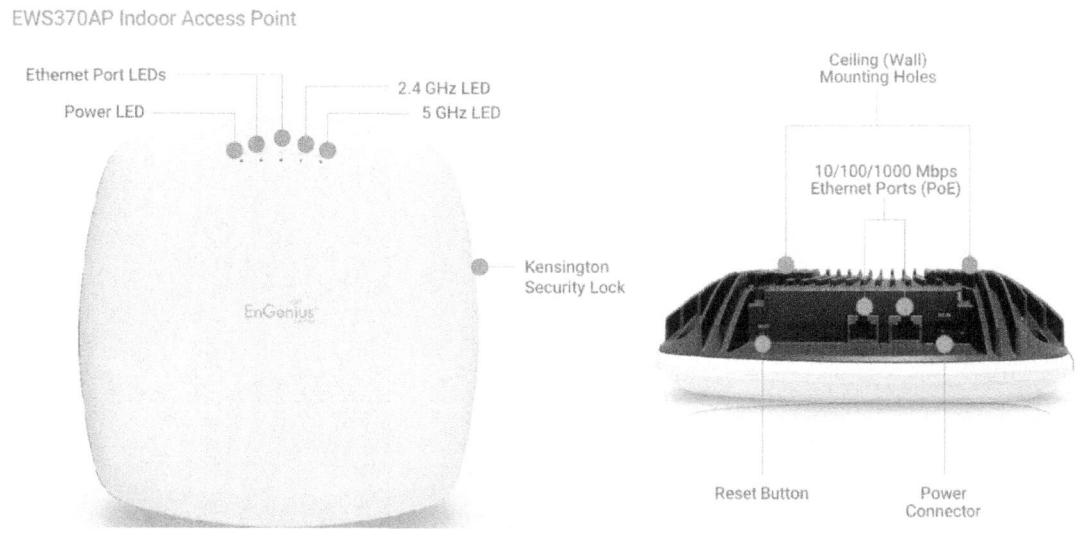

Figure 7.2: EnGenius EWS370AP Exterior Specs

The first section of a spec sheet often lists general specifications. Figure 7.3 shows this section for the EWS370AP. From here, you can determine the PHYs that are supported as well as the device memory, antenna options, interfaces, physical security options, available LEDs, power options and mounting options.

From Figure 7.3, you can see that the supported 2.4 GHz PHYs include the following:

- 802.11b (HR/DSSS)
- 802.11g (ERP)
- 802.11n (HT)

While it does not explicitly list 802.11-prime or DSSS, it is supported as well through standards-based HR/DSSS support—that is, a standards-based device that supports HR/DSSS will also support DSSS, though such support may be disabled.

The PHYs supported in the 5 GHz band include the following:

- 802.11a (OFDM)
- 802.11n (HT)
- 802.11ac (VHT)

EnGenius EWS370AP Specifications	
Standards	IEEE 802.11b/g/n on 2.4 GHz IEEE 802.11a/n/ac on 5 GHz IEEE 802.3at IEEE 802.1r IEEE 802.1k IEEE 802.11x
Memory Capacity	DDRIII: 1 GB ROM: 32 MB NAND Flash Memory 512 MB
Antenna	Internal Omni-Directional Antennas 3 dBi on 2.4 GHz & 5 GHz
Physical Interface	2 x 10/100/1000 Gigabit Ethernet Ports (Link Aggregation achieves 2Gbps Throughput) - LAN1: Supports 802.3at PoE Input - LAN2: Data Pass Through 1 x Reset Button 1 x Power Connector
Physical Security	Kensington Security Slot
LED Indicators	1 x Power 2 x WLAN (Wireless Connection) 2.4GHz 5GHz 2 x LAN
Power Source	External Power Adapter DC IN, 12V/ 2A IEEE 802.3at Compliant Source Active Ethernet (PoE) Redundant 2nd LAN Port Power
Mounting	Ceiling Mount Wall Mount

Figure 7.3: General Specs for EWS370AP

In addition, in the Standards section, you will notice a common characteristic of spec sheets. They often require interpretation and understanding of Wi-Fi. The spec sheet states that the device supports 802.1r and 802.1k, which is an actual typographic error; it actually supports 802.11r and 802.11k for roaming support. Additionally, the device supports 802.1X for security and not 802.11x. This spec sheet was not chosen for these errors, but rather it exemplifies that typographic errors are common in spec sheets. You, as the CWS, should be able to determine the intention of the stated features and capabilities.

The memory capacity of the AP is 1 GB RAM, which is sufficient to support many users concurrently using the enhanced features of the system. It also includes 32 MB ROM for the AP firmware and 512 MB NAND Flash Memory for configuration settings and possibly some firmware storage.

The next important defined specification is the internal omnidirectional antennas. Note that the antennas are 3 dBi gain antennas for both 2.4 GHz and 5 GHz. Given the shorter wavelengths in the 5 GHz band, this consistent antenna gain for both bands will require the engineer to configure output power settings appropriately if a similar cell size for the 5 GHz and 2.4 GHz bands are desired. Alternatively, in deployment, the 2.4 GHz radio could be disabled in some of the APs to address the larger coverage area typically provided by 2.4 GHz over 5 GHz.

Because the AP uses internal antennas and has no external antenna connectors, the coverage provided by those internal antennas must be accepted. It is common to use such APs for indoor deployments, where internal antennas usually work well. For warehouses and outdoor deployments as well as some other use cases, APs with external antennas are often selected instead. The external antennas provide better control over RF propagation as you learned in previous chapters.

The spec sheet next lists the physical interfaces. In this case, 2-gigabit ports are available. They are divided into a single PoE gigabit port and a non-PoE gigabit port. The PoE port can power the AP when connected to a PoE switch port or a PoE injector placed between the switch and the AP. The non-PoE port can be used as a backup data port or coupled in some way to provide an aggregate of 2 Gbps throughput on the wired side, and it can act as a redundant port for power per the later Power Source section of the spec sheet. Additionally, a wall outlet power connector is provided as well as a reset button to reset the AP to factory settings.

The Physical Security section indicates the availability of a standard Kensington Security Slot, which was also mentioned in the graphical external case specification. Such a slot is used to connect a security cable that will fasten the AP to some secured bar or panel to avoid theft. To steal the AP, the thief would have to cut the cable, which takes more time, so it acts as a theft deterrent.

The LEF Indicators section covers the same indicators referenced in the external case specification and includes a power LED, 2.4 GHz WLAN LED, 5 GHz WLAN LED and 2 LAN LEDs.

The Power Source section details the available methods for powering the AP. These options include wall outlet power through a 12-volt 2 amp DC power plug or PoE power. Notice that the specification calls for 802.3at-compliant PoE power. PoE, as used in WLANs, comes in two basic forms: 802.3af and 802.3at. 802.3af PoE provides up to 12.95 watts of power to the powered device, while 802.3at provides up to 25.5 watts of power to the powered device. This device requires the increased power provided by 802.3at, and engineers must plan for this in their switch power budget.

PoE switches provide a limited total amount of power for powered devices. While the powered device may receive 25.5 watts of power, the switch (called the Power Source Equipment [PSE]) must output 30 watts of power. Therefore, if a switch has a power budget of 150 watts, it can only power five APs demanding 25.5 watts of received power due to the fact that each AP actually requires 30 watts of output at the switch. The reason the received power is less than the output power is due to attenuation as the power travels through the Ethernet cables to the AP.

> Remember that 802.3af PoE outputs 15.4 watts of power and provides 12.95 watts of power to the powered device. Also, 802.3at outputs 30 watts of power and provides 25.5 watts of power to the powered device.

The final section of this general portion of the spec sheet shown in Figure 7.3 is the Mounting section. This section indicates support for a ceiling or wall mount. Mounting is accomplished by mounting holes on the AP. Screws are anchored into the wall or ceiling, and the holes are aligned with the screws. A slight nudge may be necessary after placing the screw heads into the holds (nudging up or down or left or right), and the AP can be mounted into place. Some APs also offer mounting kits that are first attached to the wall or ceiling, and then the APs are connected to the kit in a similar way.

The next section that is common to all wireless AP spec sheets is the wireless specifications section. The EWS370AP spec sheet calls this section Wireless & Radio Specifications and is shown in Figure 7.4. The first section is the Operating Frequency section, and it verifies that the AP is dial-radio concurrent for 2.4 GHz and 5 GHz. This statement is the same as saying that the AP is dual-band concurrent, meaning that it can operate in both bands simultaneously.

Wireless & Radio Specifications	
Operating Frequency	Dual-Radio Concurrent 2.4GHz & 5GHz
Operation Modes	Access Point
Frequency Radio	2.4GHz: 2400MHz ~ 2835MHz 5GHz: 5150MHz ~ 5250MHz, 5470 ~ 5725MHz, 5725MHz ~ 5850MHz
Transmit Power	Up to 27 dBm on 2.4 GHz Up to 27 dBm on 5 GHz Max transmit power is limited by regulatory power
Radio Chains/Spatial Streams	4x4:4
SU-MIMO	Four (4) Spatial Stream SU-MIMO up to 1733 Mbps to a single client
MU-MIMO	Three (3) Spatial Stream MU-MIMO up to 1300 Mbps to three (3) MU-MIMO capable wireless devices simultaneously
Supported Data Rates (Mbps)	2.4 GHz: Max 800 5 GHz: Max 1733 802.11b: 1, 2, 5.5, 11 802.11g: 6, 9, 12, 18, 24, 36, 48, 54 802.11n: 6.5 to 600 (MCS0 to MCS23) 802.11ac: 6.5 to 1733 (MCS0 to MCS9, NSS=1~4)
Supported Radio Technology	802.11b: Direct-Sequence Spread-Spectrum (DSSS) 802.11a/g/n/ac: Orthogonal frequency-division Multiplexing (OFDM)
Channelization	802.11ac supports very high throughput (VHT) – VHT 20/60/80 MHz channel width 802.11n supports high throughput (HT) – HT 20/40 MHZ channel width 802.11n supports very high throughput under the 2.4 GHz radio – VHT 40 MHz (256-QAM) 802.11n/ac packet aggregation: AMPDU, ASPDU
Supported Modulation	802.11b: BPSK, QPSK, CCK 802.11a/g/n: BPSK, QPSK, 16-QAM, 64-QAM 802.11ac: BPSK, QPSK, 16-QAM, 64-QAM, 256-QAM

Figure 7.4: Wireless & Radio Specifications Spec Sheet Section

The next section indicates operating modes. In this case, only one mode is offered: Access Point. In addition to the AP operating mode, some APs can operate in bridge mode. Bridge mode allows the AP to form a bridge link with another AP to create longer distance links to bridge otherwise separate networks. An additional mode that is sometimes offered is the repeater mode. In the repeater mode, the AP is a client to another AP, and it also serves the role of AP to remote clients. This can be used to allow distant clients to connect to the network; however, repeater mode is not recommended, since all frames received from clients must be repeated from the repeater AP to the other connected AP. Additionally, when the repeater AP receives a frame from the other connected AP for a client, it must retransmit that frame. Therefore, the throughput is cut in half in this operating mode.

The next section, Frequency Radio, indicates the operational frequencies that are supported by the AP. In the 2.4 GHz band, this is nearly always the entirety of the 2.4 GHz band that is usable by 802.11 devices. The important area to inspect here is the 5 GHz support. Notice that this AP supports 5150 MHz to 5250 MHz (lower range defined in Chapter 4), 5470 MHz to 5725 MHz (middle range defined in Chapter 4) and 5735 MHz to 5850 MHz (upper range defined in Chapter 4). Therefore, the AP can support all available 5 GHz channels. In most cases, APs will support all of the 5 GHz channels, with some exceptions for channels 155 and 165, but it is important to verify this.

The next section, Transmit Power, simply lists the maximum potential transmit power for each band. In both bands, it is stated as 27 dBm, which translates to 500 mW of output power. However, this is constrained by regulatory domains. When the AP is configured with a country code that specifies the regulatory domain, it defaults to the maximum transmit power allowed in that domain. In fact, in the configuration interface, you cannot reduce this power by default. Figure 7.5 shows the configuration interface for the power settings of the EWS370AP. Notice that the Transmit Power setting, which controls the output power of each radio, is disabled. To enable it, you must deselect the "Green" checkbox in the upper right corner, and then you can setup the output power. It is important to use this setting for output power control with care because the AP can be configured with output power settings that are disallowed in your regulatory domain.

The next section lists the radio chains and spatial streams that are supported. You can see the common nomenclature that was introduced in Chapter 6. The AP supports 4x4:4 radio chains and spatial streams with an end result of support for 4 spatial streams. Remember that the actual spatial streams used will depend on the lowest common denominator, which is usually the client. Given that most available clients at the time of writing support only three or fewer spatial streams, the link will be limited as such. However, the EWS370AP supports Multi-User

MIMO (MU-MIMO), so the four radio chains could also be used to simultaneously transmit to two 2x2:2 clients.

	2.4GHz		5GHz	
Operation Mode	Access Point	✓ Green	Access Point	✓ Green
Wireless Mode	802.11 B/G/N		802.11 AC/N	
Channel HT Mode	20MHz		40MHz	
Channel	Configuration			
Transmit Power	Auto		Auto	
Data Rate	Auto		Auto	
RTS/CTS Threshold (1 - 2346)	2346		2346	
Client Limits	127	Enable Disable	127	Enable Disable
Aggregation	Enable Disable			
	32 Frames			
	50000 Bytes(Max)			
AP Detection	Scan		Scan	
Distance (1-30km)	1 (0.6miles)		1 (0.6miles)	

Figure 7.5: EWS370AP Output Power Configuration

The next section lists the radio chains and spatial streams that are supported. You can see the common nomenclature that was introduced in Chapter 6. The AP supports 4x4:4 radio chains and spatial streams with an end result of support for 4 spatial streams. Remember that the actual spatial streams used will depend on the lowest common denominator, which is usually the client. Given that most available clients at the time of writing support only three or fewer spatial streams, the link will be limited as such. However, the EWS370AP supports Multi-User MIMO (MU-MIMO), so the four radio chains could also be used to simultaneously transmit to two 2x2:2 clients.

> *MU-MIMO* allows an AP to transmit to multiple clients simultaneously and is a feature of the 802.11ac standard. To accomplish MU-MIMO communications, many factors must be in alignment, and it is unlikely to yield massive improvements, but it is an available feature.

The SU-MIMO and MU-MIMO sections are next on the spec sheet (Figure 7.4). As previously stated, MU-MIMO allows the AP to simultaneously transmit to multiple client stations. The EWS370AP states that it offers single user MIMO (SU-MIMO) support with up to 4 spatial streams for a maximum data rate of 1733 Mbps to a single client. This data rate would require an 80-MHz channel, so it is unlikely that a production deployment of this AP would ever have a single client operating at that data rate. However, on a 40-MHz channel, an 800-Mbps data rate could be achieved.

The spec sheet also indicates that up to 3 clients can simultaneously receive MU-MIMO transmissions. This is somewhat more challenging to understand. The spec sheet states the following: "Three (3) Spatial Stream MU-MIMO up to 1300 Mbps to three (3) MU-MIMO capable wireless devices simultaneously." Does this mean that three STAs can each receive 1300 Mbps of data for an aggregate of 3900 Mbps? In fact, it does not. The 1300 Mbps is an aggregate value, and each MU-MIMO client actually achieves an independent data rate of 433.3 Mbps. Again, the engineer should understand this, and the CWS should also understand this when recommending or selecting wireless APs. Additionally, this 433.3 Mbps depends on an 80-MHz channel. For this reason, when using 40-MHz channels, you are more likely to see an aggregate of 600 Mbps with three simultaneous MU-MIMO clients operating at 200 Mbps each. Yes, situations can become quite complex, but with a little digging into the standard (specifically the MCS parameters) and careful reading of the spec sheets, you can determine what you need to make an effective decision.

In Figure 7.4, the next section is Supported Data Rates. This spec sheet lists the maximum potential data rates as 800 Mbps for 2.4 GHz and 1733 Mbps for 5 GHz. To achieve the 800 Mbps data rate in the 2.4 GHz band, a simple visit to the 802.11 standard reveals that there is something awry. The 802.11ac (VHT) MCS tables allow for 800 Mbps with four spatial streams and a 40-MHz channel. Is this spec sheet suggesting to use VHT in the 2.4 GHz band and a 40-MHz channel? Perhaps, but this is not practical since the clients would not support it, and you should never use a 40-MHz channel in the 2.4 GHz band. Therefore, the true maximum data rate when using 2.4 GHz is based on the 802.11n (HT) PHY, which is supported with 2.4 GHz and a 20-MHz channel with three spatial streams for a data rate of 216.7 Mbps.

Confusion may arise when you see that the spec sheet states that 802.11n (HT) MCS 0-23 are supported, which do not include 4 spatial streams. So, the 800 Mbps maximum remains a bit of a mystery using the 2.4 GHz band for this AP.

Additionally, in the 5 GHz band, 80-MHz channels are not likely to be used, so the true maximum data rate for production environments will be based on four spatial streams and a 40-MHz channel. This data rate will be 800 Mbps. The additional data rates are standard for the 802.11b/g/n/ac specifications.

The Supported Radio Technology section reveals that DSSS (and therefore HR/DSSS) is supported in the 2.4 GHz band, and OFDM is supported in the 2.4 GHz and 5 GHz bands. Remember from preceding chapters that OFDM is used in ERP, OFDM, HT and VHT PHYs. This fact is what is indicated by the line that reads "802.11a/g/n/ac: Orthogonal frequency-division Multiplexing (OFDM)."

The next section, *Channelization*, indicates the supported channel widths that are used by radios. As an 802.11ac (VHT) radio, channel widths of 20, 40 and 80 MHz are supported. As an 802.11n (HT) standard radio, channel widths of 20 and 40 MHz are supported. Next, we see the answer to the dilemma of 800 Mbps as the max data rate in the 2.4 GHz band. Note the line that reads, "802.11n supports very high throughput under the 2.4 GHz radio – VHT 40 MHz (256-QAM)." This is where the 800 Mbps comes from. However, it is a misleading statement. The 802.11n standard does not support VHT 40 MHz with 256 QAM; the EnGenius implementation does support this, but the standard does not. That is, the chipset used in the EnGenius AP supports using 256 QAM modulation in the 2.4 GHz band, but this is not a standard feature. Therefore, most clients will not be able to take advantage of this enhancement.

The final section of the spec sheet depicted in Figure 7.4, Supported Modulation, simply indicates the modulation options supported by the AP. These options are standards-based and compliant with 802.11.

The remaining portions of the spec sheet (not shown in an image here) define other important features. For example, the EWS370AP can support up to 8 SSIDs per radio. This feature allows for multiple SSIDs when required. It is common, for example, to use one SSID for the enterprise users and another for guest users. However, when multiple SSIDs are used, it is important not to overuse them. Each SSID requires its own beacon frame to be transmitted approximately ten times each second. Therefore, 2 SSIDs require approximately 20 beacons per second, 3 SSIDs require 30 beacons per second and so on. These extra beacons result in the higher overhead on the wireless medium and reduce overall throughput. Most engineers recommend using three or fewer SSIDs per radio.

Another important specification is the management interface options for an AP. The EnGenius AP, which was used as an example here, can be managed through a Web interface or Telnet (through SSH is preferred). When managing an AP directly through Web interfaces, HTTPS should always be used as it encrypts communications, unlike HTTP. For the same reason, SSH should be used and not simple Telnet, because SSH encrypts the communications.

The EWS370AP can also be managed through a central switch that remotely manages the AP and provides automatic discovery and provisioning of the AP, automatic IP address assignment and overall group management so that multiple APs may be quickly deployed.

An additional specification often sought in AP solutions is support for guest networks. Some autonomous (stand-alone APs) solutions support guest networks, and others require a controller-based infrastructure or some other managed infrastructure. These management solutions are addressed in the next section of this chapter. To offer guest network access, most supporting APs implement a separate SSID for guest access and may also implement a captive portal for authentication or use a policy agreement.

Translating marketing speak or language into real-world deployments is an important skill for a CWS or any wireless professional. When you read of maximum data rates, always remember that the client dictates the actual performance.

Finally, a good spec sheet will identify the security solutions supported by the AP. For example, the EnGenius EWS370AP lists the following supported security features:

- WPA/WPA2 Enterprise (PSK is supported as well)
- Hide SSID in beacons (which is not a security feature, but it is listed as such)
- MAC address filtering (which is not a security feature, but it is listed as such)
- HTTPS support
- SSH support
- Client isolation (which prevents BSS STAs from communicating directly with each other)

> Any negative comments in this section related to the EWS370AP spec sheet are not intended to indicate some uniqueness of this spec sheet. Today, all vendors provide information on spec sheets that can be misleading or at best confusing. This is an area that requires much improvement.

AP Management Solutions

The final section of this chapter is focused on AP management. APs can be managed in two primary ways: individually or in groups. When managed individually, they are configured as autonomous or stand-alone APs. When managed in groups, they are managed by controllers, the cloud or some other management solution. All of these methods are explained here.

Autonomous

Autonomous APs are individually managed. They are configured through Web-based interfaces or command line interfaces (through SSH). Autonomous APs may be configured through a centralized management system, sometimes called a Wireless Network Management System (WNMS), but these are less common today as independent solutions. Most WLAN vendors provide centralized methods for configuring even autonomous APs without requiring third-party solutions. This is true for most enterprise vendors.

The advantage of autonomous APs is that they have no dependency on some other component to continue operations—that is, the AP works alone and does not require something like a controller or a cloud system to continue operations. However, many controller- and cloud-based APs today also offer some form of continuity of operations even when the controller or cloud is unavailable.

The disadvantage of autonomous APs is that they must be managed individually. For this reason, autonomous APs are useful for smaller implementations, but they do not scale well. Some form of centralized management is required when dozens or hundreds of APs are deployed.

Controller

Controller-based APs are managed by a centralized device or a virtual controller that runs in a virtual machine (VM) or in an AP itself. Many vendors now support virtual controllers in one of the APs—that is, one of the APs will act as the controller for the other APs. Virtual controllers are an efficient solution for smaller deployments. They do not scale well since the APs lack the necessary processing power to handle hundreds or thousands of other APs.

When using a controller, two primary options exist for data forwarding onto the wired network: centralized and distributed forwarding. Centralized forwarding sends all data back to the controller, and the controller then forwards the data to the destination on the wired network. In the other direction, data destined for wireless clients go through the controller to the AP and then to the appropriate client. Some have suggested that centralized forwarding is a significant bottleneck, but in many infrastructures, it works well with very little added delay.

Distributed forwarding allows data to be transmitted from the APs directly to the wired network. To understand how this works, it is useful to know that there are three planes of network traffic on wireless networks: the control plane, the management plane, and the data plane. The control plane is used for things like automatic channel assignment and infrastructure-based roaming. The management plane is used for things like AP configuration and monitoring. The data plane is for user data.

With this explanation of traffic planes, you can see that the data plane can be forwarded, while control and management plane information can pass through the controller. For the most part, management and control plane information are all that has to be passed to the controller. With distributed forwarding, more latency is removed from the transmission of data frames to and from wireless clients.

> The *control plane* is about control of network access and RF signaling. The *management plane* is about device management. The *data plane* is about user data transmission. Remember these three divisions, and you will understand a great deal of vendor literature.

Cloud

Cloud-based APs are managed and monitored by a cloud solution. Vendors like Aerohive, Meraki, and others offer such cloud solutions. They differ greatly in the specifics of control and management plane information, but they all support a form of distributed data forwarding for the local network.

Some vendors implement what they call a cooperative control architecture. In this architecture, the APs maintain the control, management and data planes of all Wi-Fi communications, but the management of the APs is centralized in the cloud. Additionally, APs are monitored by the cloud for management, troubleshooting and maintenance purposes.

Other vendors do not claim cooperative control architectures but offer very similar end results. When looking for a cloud-based AP management solution, the most important factor is to ensure that the APs continue to operate even if the cloud is unreachable. If they cease to function when the cloud is unavailable, your WLAN will stop working, and your users will not be happy.

An additional concern with cloud management solutions is data privacy. Today, most cloud vendors do not pass user data to the cloud to avoid privacy concerns with the actual data. However, they do track the applications used and sometimes the websites accessed in their monitoring solutions. If this is a privacy concern to your organization, you must understand options for disabling such tracking or select a vendor that does not perform this tracking.

> Cloud-managed APs are much like WNMS-managed APs of the past, except the management system is in the cloud. You can configure, monitor and push firmware to APs from the cloud in the same way you did from older WNMS systems.

Management Systems

Finally, the general category of management systems is inclusive of all other management solutions. For example, some vendors offer management solutions that are similar to the cloud but are installed on the local network through servers, appliances or wireless management switches.

An example of a management system that does not neatly fit into the model of controller or cloud (strictly speaking) systems is the Ubiquiti UniFi Controller Software. This software can be installed on a local server to manage the Ubiquiti APs, or it can be installed using a UniFi Cloud Key, which is a very small device that includes the controller software on it.

Whatever the chosen management system, it is essential to understand the organizational requirements and select both APs and management solutions that meet these requirements.

Chapter Summary

In this chapter, you learned about APs and their features and capabilities by exploring a sample spec sheet. You also explored AP management options. In the next chapter, you will learn about the wireless clients that connect to these APs and management systems.

Points to Remember

Remember the following important points:

- A specifications sheet (spec sheet) provides detailed information about device features and capabilities, and APs typically have spec sheets.

- Spec sheets or data sheets often provide external views of the AP with details about LEDs, ports and mounting options.

- A dual-band concurrent AP is one that can run in the 2.4 GHz and 5 GHz bands at the same time.

- Most APs support all older PHYs as well as the stated AP PHY (such as an 802.11ac AP) in the same band for backward compatibility.

- Internal antennas are often used for standard office deployments, but external antennas are often desired in special use cases, such as warehouses, outdoor deployments, and large venues.

- When PoE is available for an AP, it is important to know what version of PoE is required: 802.3af (12.95 watts delivered to the powered device) or 802.3at (25.5 watts delivered to the powered device).

- APs often support three operating modes: Access Point, Bridge, and Repeater.

- Output power settings should be configured in relation to the regulatory constraints in which the AP operates.

- Identifying the radio chains supported is a key factor in selecting APs.

- Translating marketing language into real-world deployments is a skillset that the CWS candidate should develop.

- Autonomous APs are managed individually.

- Controller-based APs may use centralized data forwarding or distributed data forwarding.

- Cloud-based APs are typically configured (as well as monitored and managed) from the cloud, but operations can continue if the cloud is unavailable.

Review Questions

1. What document may be used to learn about the details of a specific AP model?
    a. IEEE 802.11-2016
    b. Spec sheet
    c. Architectural whitepaper
    d. IETF RFC 3680

2. What external component of an AP may be documented in a spec sheet?
    a. Filters
    b. Chipset
    c. LEDs
    d. Modulation

3. When an AP can simultaneously operate in the 2.4 GHz and 5 GHz bands, what is this called?
    a. DRS
    b. SISO
    c. MIMO
    d. Dual-band concurrent

4. Where are external antennas most likely to be used?
    a. Office space
    b. Conference room
    c. Warehouse
    d. Front lobby

5. What is the power provided to a powered device when 802.3af is used?
    a. 5 watts
    b. 12.95 watts
    c. 15.4 watts
    d. 25.5 watts

6. What is the power provided to a powered device when 802.3at is used?
    a. 12.95 watts
    b. 15.4 watts
    c. 25.5 watts
    d. 30 watts

7. What specific technology allows an AP to transmit to more than one client at the same time?
    a. MIMO
    b. SISO
    c. SU-MIMO
    d. MU-MIMO

8. What skill set is required to understand how the details in a spec sheet may apply to real-world deployments?
    a. Use of a spectrum analyzer
    b. Use of a site survey tool
    c. Translation of marketing language
    d. Memorizing the 802.11-2016 standard

9. What protocol should not be used to manage an AP?
    a. HTTP
    b. HTTPS
    c. SSH
    d. None of these

10. What solution is used to allow controller-based APs to send data directly to the destination on the wired network?
    a. Centralized forwarding
    b. Distributed forwarding
    c. Call-back forwarding
    d. Routed forwarding

Review Answers

1. **B is correct.** The specifications sheet (i.e., spec sheet) or data sheet may be used to gather details about the features and capabilities of an AP model.

2. **C is correct.** External components such as LEDs, wired ports, reset buttons and mounting options are often defined.

3. **D is correct.** A dual-band concurrent AP is one that can operate in the 2.4 GHz and 5 GHz bands at the same time.

4. **C is correct.** External antennas are likely to be used in warehouses, manufacturing plants, outdoor deployments, and high-density scenarios.

5. **B is correct.** With 802.3af, the output power from the PSE is 15.4 watts, but the power provided to the powered device is 12.95 watts.

6. **C is correct.** With 802.3at, the output power from the PSE is 30 watts, but the power provided to the powered device is 25.5 watts.

7. **D is correct.** MIMO is used, but the specific technology is multi-user MIMO (MU-MIMO).

8. **C is correct.** All wireless professionals must know how to translate marketing language into real-world deployment scenarios.

9. **A is correct.** Protocols that send information as clear text (e.g., HTTP, Telnet) should not be used for AP management.

10. **B is correct.** Distributed data forwarding, or simply distributed forwarding, allows APs to send data directly to the destination on the network.

Chapter 8 – Wireless clients

Objectives Covered

3.3 Determine capabilities of client devices
    3.3.1   PHY support
    3.3.2   Single-band vs. multi-band
    3.3.3   Support for MIMO
    3.3.4   Supported channels in 5 GHz
    3.3.5   Supported security options

4.4 Discover client devices and applications in use
    4.4.1   Laptops, tablets, mobile phones, desktops and specialty devices
    4.4.2   Real-time applications
    4.4.3   Standard applications (e-mail, web browsing, database access, etc.)
    4.4.4   Data-intensive applications (file downloads/uploads, cloud storage, cloud backup, etc.)

As you learned in previous chapters, the client devices on a WLAN are very important. In many cases, they determine the actual capabilities of the network. This statement is true because of the law of the least common denominator. Remember that in wireless networking, within a link, the least common denominator determines the available features. For this reason, understanding client device types and the features of the various types is essential. This chapter provides details on important considerations for the clients on your WLAN. With the information provided here, you will be better prepared for the CWS exam and for selecting or recommending the right WLAN solutions.

Client Device Types

WLAN client devices come in many shapes and sizes and are targeted for multiple purposes. The traditional laptop is still used, but in the grand scheme of things, it is probably the least used mobile device on many WLANs. On public hotspots, they are certainly a minority, and given the BYOD movement, many enterprises must handle the load of mobile phones and tablets. In this section, the basic characteristics of these devices will be explained.

Laptops

Laptop computers are basically portable desktops in that they run the same operating systems as desktop computers. The majority of laptops today run Windows, macOS or Chrome OS. A smaller subset run a distribution of Linux. Because they run these operating systems, the default support for security features is well known. For example, in Windows, you can run PEAP, EAP-TLS, EAP-TTLS, and EAP-SIM as the EAP method for WPA2-Enterprise authentication. In macOS, you can run PEAP, EAP-TLS, EAP-TTLS, EAP-FAST, and EAP-SIM. In Chrome OS, you can run PEAP, EAP-TLS, and EAP-TTLS. Finally, the default WPA supplicant in Linux systems supports all of the listed EAP types and more. The point is simply that you can standardize on PEAP, EAP-TLS or EAP-TTLS, and nearly any laptop will be able to connect to the network.

> The compatibility with given EAP methods varies by operating system version as well. You should always check with the operating system vendor to verify that the desired EAP method is supported in your version of the operating system.

The available wireless hardware in laptops varies considerably. Some laptops are single-stream 802.11n devices that only operate in the 2.4 GHz band. Others are dual-band, three-stream 802.11ac devices. When planning a WLAN, it is important to define which clients will be used within the organization and ensure the best WLAN is implemented to support these clients. In most cases, the least capable devices on the network are the most important to the final design.

> If you are able to influence purchasing decisions, 2.4 GHz-only devices should no longer be purchased. A 5 GHz WLAN can easily be configured to outperform a 2.4 GHz WLAN. Having 5 GHz-capable clients is always best.

In most cases, the internal wireless NIC in laptops is an adapter card that may be replaced. Some laptops (e.g., macOS or Chrome OS laptops) may offer fewer adapter replacement options. Windows-based laptops are often upgradable through replacement of mini-PCIe adapters or other adapter types. When replacing them, it is important to ensure the new internal adapter supports the same number of antennas as the old adapter because you will be limited to that number of antenna connections in the laptop design.

Figure 8.1 shows a mini-PCIe adapter of the type that is often used within laptops. This pictured adapter is an AzureWave Broadcom BCM94352HMB/BCM94352 802.11ac adapter and is a half mini-PCIe adapter. It is dual-band (though clients only work in one band at a time from a connection perspective) and supports Bluetooth as well as Wi-Fi.

Figure 8.1: AzureWave Broadcom BCM94352HMB/BCM94352 Half Mini-PCIe Adapter

An additional form factor for internal wireless adapters that is becoming more common was originally called Next Generation Form Factor (NGFF), but more recently it has taken on the formal name of M.2. This form factor is similar to mini-PCIe, and many devices are released in both forms. For example, the Tri-Band Qualcomm solution, which is referenced in the *Single-Band vs. Multi-Band* section later in this chapter (chip QCA9006), comes in both the NGFF and half mini-PCIe form factors.

Laptops may also use external USB 2.0 or 3.0 adapters, which are very common today. They range from 802.11n, which is still on the market, to 3x3:3 802.11ac adapters. Such adapters are often used by WLAN engineers in the process of designing wireless networks. They work well with site survey software and protocol analysis software. Legacy devices may use Cardbus/ExpressCard adapters, but the market has moved mostly toward USB.

Figure 8.2 shows an Edimax EW-7833UAC USB 3.0 adapter. This adapter works with some protocol analyzers (e.g., CommView for Wi-Fi) and is a three-stream USB adapter.

Tablets and Mobile Phones

Tablets and mobile phones have a lot in common. They often use similar processors and run similar operating systems. For example, iPhones and iPads both run iOS. Android phones and tablets both run the Android OS. Windows phones and some tablets use the mobile version of Windows. Most tablets and phones support PEAP, EAP-TLS, and EAP-TTLS, and most mobile phones also support EAP-SIM. Where they differ is in their receiving sensitivity and wireless PHY support.

Many tablets and mobile phones that are in use today still use one- or two-stream 802.11n wireless chips. Some newer devices use one- or two-stream 802.11ac chips. It is important to understand the limitations these devices may bring to your network. It is likely that the mobile phones and tablets will have slower wireless adapters compared to laptops and desktops. They will slow the network down because they consume more airtime to transmit the same data.

Additionally, you must consider these mobile device types from an access perspective. Many are owned by the users and not the enterprise. BYOD solutions discussed in previous chapters may be implemented to onboard and manage these devices.

The following list is a sampling of capabilities of various mobile tablets and phones used on enterprise networks:

- Amazon Kindle Fire HD – 802.11n, 1 spatial stream
- HTC One (M8) – 802.11ac, 1 spatial stream
- iPad Pro – 802.11ac, 2 spatial streams
- iPad Air 2 – 802.11ac, 2 spatial streams
- iPhone 5s – 802.11n, 1 spatial stream
- iPhone 6 – 802.11ac, 1 spatial stream
- iPhone 7 – 802.11ac, 2 spatial streams
- LG G4 – 802.11ac, 1 spatial stream
- Microsoft Lumia 950 – 802.11ac, 2 spatial streams

Figure 8.2: Edimax 3-Stream USB 3.0 Adapter

From this list, you can see that no current (at the time of writing) tablet or mobile phone supports three spatial streams – even with 802.11ac. This decision makes sense when you consider how the devices are used. The extra spatial stream would drain the battery faster without adding significant value. In fact, it is questionable whether these devices even need two streams at this time.

Desktops

Desktops should be connected to wired connections when they are available. It is that simple. However, use cases exist that demand the use of wireless with a desktop. For example, a desktop placed in a remote corner of a warehouse that has wireless coverage but no Ethernet outlet can benefit from a wireless connection. Desktops can use internal PCI 802.11 NICs or USB-based NICS. The benefit of an internal PCI NIC is that it is not easily unplugged like a USB device. It is not uncommon for a user to unplug the USB wireless adapter and then wonder why the network connection is not working. The negative aspect of an internal PCI adapter is that the antennas are often near the floor, behind the computer case, and under the desk. This placement is not optimal for wireless performance. Some PCI adapters exist that offer external antennas with cable runs that reach the top of the desk, and these tend to perform better in such scenarios.

A USB adapter offers the flexibility of easy replacement, in spite of the "user unplugged it" issue previously mentioned. You can easily replace the adapter with a newer one to offer faster PHY capabilities as they become available.

> When selecting a USB or PCI wireless adapter, select one that is compatible with your operating system. Also, it is often best to download the newest drivers to enable all features. Some USB adapters literally ship with drivers that do not allow the use of the 5 GHz band.

Specialty Devices

Specialty devices are devices that use 802.11 networking but are not traditional user devices like laptops, desktops, tablets or mobile phones. They include push-to-talk devices, health monitoring devices, alarm systems, video surveillance systems and even door locks.

Figure 8.3 shows a wireless door lock. This particular lock, and sadly most such devices, is a 2.4 GHz-only device. Even worse, these devices are typically 802.11b/g devices. Therefore, once installed (and likely to be used for 5-7 or more years), it will continue to significantly slow the 2.4 GHz-band down. In many office spaces, such locks will be part of every single BSS in the building, resulting in poor performance across the board. Thankfully, they do not transmit a significant amount of data, but their existence ensures that compatibility methods will have to be used for a long time to come.

> You may wonder why devices like door locks and others still use older 802.11b/g chipsets. The answer is twofold: price and development costs. Older radios are cheaper, and continuing to use older devices means that new development for new radios does not have to be performed.

Figure 8.4 shows a push-to-talk digital badge. It works as both an employee ID device and a communication device. Such devices are common in healthcare deployments. This particular device, the Zebra Enterprise SB1 Smart Badge by

Motorola, is an 802.11b/d/n device that supports push-to-talk VoIP, barcode scanning and distributed task list management all in one device. Again, this is an example of a 2.4 GHz-only device, which introduces challenges to providing good voice quality because the 2.4 GHz band is so congested.

Figure 8.3: Wireless Door Lock (Schlage NDE)

Figure 8.4: Zebra Enterprise SB1 Push-to-Talk Badge (Motorola)

Single-Band vs. Multi-Band

Client devices may be single-band or multi-band. Notice that the term is not dual-band. The reason for this is simple: some client devices now support 802.11n for 2.4

GHz, 802.11ac for 5 GHz and 802.11ad for 60 GHz. The latter are rare at the time of this writing, but they do exist. The Acer TravelMate P648 shown in Figure 8.5 is an example of such a device. The laptop uses the Qualcomm Tri-Band Atheros chip (QCA9006), which supports all three bands. Given that 802.11ah operates in the Sub-1-GHz band, it is possible we will see Quad-Band chips as well. For this reason, CWNP has chosen to standardize by using the term multi-band.

Figure 8.5: Acer TravelMate P648 Laptop (802.11n, 802.11ac and 802.11ad)

In most cases, multi-band client devices can use only one band at a time (see the next note for real-world implications of this fact). The exception to this statement is when multiple adapters are used in the client device, which means that for practical purposes, only laptops may have an exception. The practical need for multi-band concurrency is not there for more applications, so this is rarely ever an issue.

> It is for this reason that statements like "speeds up to" followed by an aggregate data rate for both bands can be misleading to consumers. They expect a single client to achieve those speeds, which never happens in the real world.

Some devices only work with the 2.4 GHz band, and such devices should be avoided whenever possible. Very rarely, a device may only work with the 5 GHz band, which is sometimes a problem as well, because the device may need to connect to the far more popular 2.4 GHz band network as well. It is not uncommon to encounter 2.4 GHz-only coverage in some areas. This yields a poor performance result, but it is very common. If the network is not redesigned to allow for 5 GHz connectivity, a 2.4 GHz radio will be required. Ideally, the network should be redeployed to support all 5 GHz clients everywhere instead of forcing them onto 2.4 GHz.

> In fairness, some deployments require only 2.4 GHz, but they are rare today. An example would be a warehouse that uses 2.4 GHz scanners and has no other need for Wi-Fi whatsoever. This scenario would warrant a 2.4 GHz-only network.

Physical Layers (PHY) Supported

Ultimately, part of being a multi-band device is indicative of PHYs supported to some extent. For example, if a device supports both 2.4 GHz and 5 GHz, it must support the 802.11 (DSSS) PHY and 802.11a (OFDM) PHY at a minimum. However, that is the minimum it must support. Most devices that support the 2.4 GHz PHY will support at least 802.11b (HR/DSSS), and it is very common for the minimum supported PHY in an organization's devices to be 802.11g (ERP). Today, the goal should be to use 802.11n (HT) as the minimally supported PHY in the 2.4 GHz band.

In the 5 GHz band, the absolute minimally supported PHY is 802.11a (OFDM), but the goal should be to support 802.11n (HT) or greater (meaning 802.11ac [VHT] at the time of writing, though 802.11ax is coming soon).

When purchasing client devices today, the decision maker should evaluate the PHYs that are supported by the device. If a device meeting the needs of the project can be acquired that supports 802.11n in the 2.4 GHz band and 802.11ac in the 5 GHz band, that device should be preferred. The next option would be a device that supports 802.11n in both bands. Sadly, some device types simply cannot be acquired with chips that support anything greater than 802.11g and the 2.4 GHz band (e.g., door locks). This reality needs to quickly change in the industry.

> To obtain speeds advertised by an AP spec sheet, clients must support the highest data rate PHYs that the AP supports. They must also support the highest number of spatial streams that the AP supports. Otherwise, real-world data rates will be less than the AP's capabilities.

It is also important to know that the actual speeds achieved in the network will usually be driven by the combination of devices used. Even when a device connects at a very high data rate (using the 802.11ac PHY, for example), it will be impacted by the slower devices on the network that need airtime as well. The end result will be that the throughput of the faster device will be much lower than its potential.

Supported Channels

Generally speaking, client devices designed for sale in a given regulatory domain may support all of the allowed 2.4 GHz channels in that domain. However, it is not uncommon for 2.4 GHz devices sold worldwide to constrain support to channels 1 through 11 simply because those channels are supported everywhere in the world. This fact makes 2.4 GHz channel support an easy factor to understand.

When it comes to 5 GHz client devices, the story is much different. According to Mike Albano's tracking, which is reported at clients.mikealbano.com (shown in Figure 8.6), more than 60% of tracked devices do not support channel 144. This should be expected, since many of the devices are 802.11n, and channel 144 was added with 802.11ac. However, a portion of the devices that do not support channel 144 is 802.11ac devices. This is important to know, and you need to understand the client device pool in a deployment to decide whether this channel should be used or not.

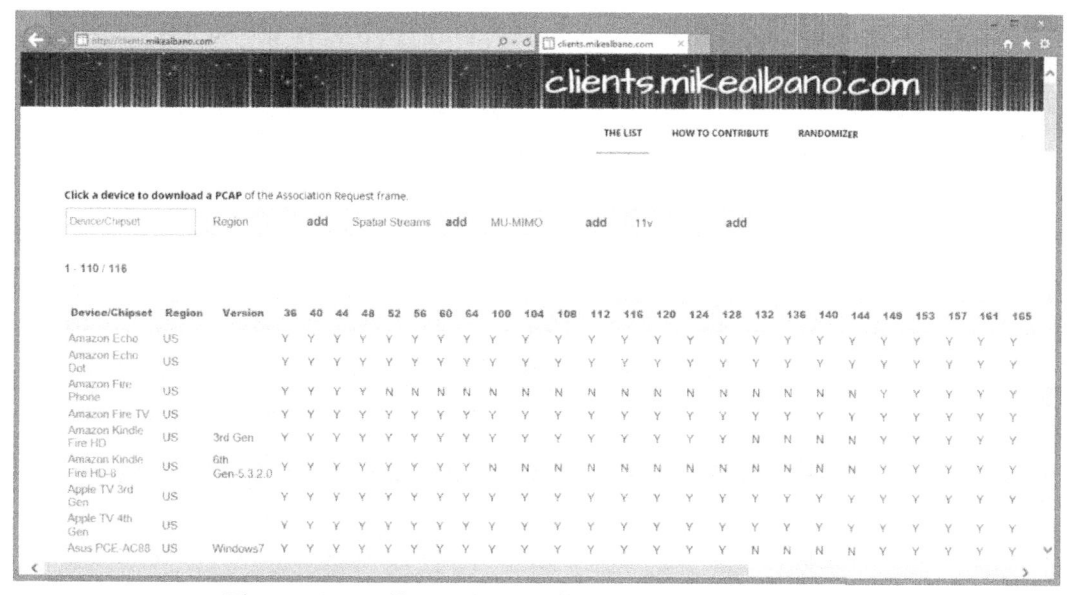

Figure 8.6: Client List at clients.mikealbano.com

> When evaluating client channel support, pay close attention to the 5 GHz band. This band is more likely to see unsupported channels than the 2.4 GHz band. This fact is particularly true in the middle range of 5 GHz channels.

Of even greater significance is the number of devices that do not support some or all of the middle range of channels (i.e., channels 100 to 144). These devices make up more than 16 percent of the evaluated pool. Some of these devices also lack support for channels 52 through 64, making the issue even more challenging. In the end, channels 36 through 48 and 149 through 165 are very safe. These channels provide nine 20-MHz channels that can provide a first layer of coverage throughout a deployment. Additional APs can be deployed on the other channels to provide an extra layer of coverage and capacity. Using this strategy allows you to support all client types; however, there is not a consistent way to force clients that support the middle range of channels to use those APs, which would then leave the other APs for the clients that do not support the middle range. As you can see, the complexity of the design can be associated with the 5 GHz band just like the 2.4 GHz band; they simply represent different challenges.

Supported Security Options

The next factor in client device documentation is the supported security options. These options include support for WPA or WPA2 and the EAP methods supported by the devices.

Some older devices still only support WPA. Particularly among VoIP handsets and barcode scanners, you will still see a lot of devices in production that require WPA and do not support WPA2. The CWS should recommend that these devices be replaced or upgraded to support WPA2; however, some organizations will not be willing to make that investment because security is not a high concern for them. If all attempts fail to convince them to upgrade, you will need to make them aware of the following:

- Using WPA instead of WPA2 is less secure.

- Enabling WPA on an SSID will disable 802.11n and 802.11ac PHY rates, and all devices that connect to the SSID will be impacted by this.

- An alternate SSID that supports WPA can be configured on the APs, but the decision maker should be informed that this configuration will reduce the performance of the WPA2 SSIDs running on the same AP because of the extra beacon frames that are required for an extra SSID.

It is always best to upgrade devices or replace them with WPA2-compatible configurations or devices. When this is not possible, an alternate SSID is an option, but it will degrade the performance of other SSIDs, or BSSs, offered in the same band by the same AP.

It is also important to document the EAP methods supported by the client device population. PEAP, EAP-TLS, and EAP-TTLS are typically very safe because most devices support them. Some security concerns have arisen over the years related to PEAP, but when implemented properly, it still offers good security. The best security comes from EAP-TLS and EAP-TTLS, which are among the popular EAP methods. However, you must remember to inform the decision maker of the certificate requirements for EAP-TLS and EAP-TTLS. EAP-TLS will require a server (the authentication server or RADIUS server) and a client certificate for each participating client device. EAP-TTLS requires the server certificate and does not require client certificates.

PEAP can be implemented in different ways, and it uses internal authentication mechanisms. If implemented with MS-CHAPv2, no certificates are required. When implemented with EAP-TLS as the internal authentication mechanism, server and client certificates are required.

Applications in Use

The final element to consider with clients is the application pool that they intend to utilize. Client applications range from those requiring little of the network to those with specific demands on latency and reliable delivery. The following states the information that should be gathered about each client application:

- **Latency tolerance:** Do the packets have to arrive in some constrained amount of time? For example, VoIP packets should arrive in less than 150 ms.

- **Throughput requirements:** How much data will be passed through the network by the application? For example, 100 Kbps, 512 Kbps, 1 Mbps and so on.

- **Utilization levels:** How often will the application communicate on the network? Some applications will use the network less than 1 percent of the time, and others may communicate up to 40 or 50 percent of the time with peaks where communications happen for nearly 100 percent of the time for some period.

- **Protocols used:** Does the application use TCP or UDP? TCP requires a connection and provides reliable delivery with slightly more overhead. UDP does not typically require a connection and does not offer reliable delivery (meaning that delivery is not usually confirmed).

- **Ports used:** What TCP or UDP ports are used by the application? This information is important for the configuration of routers and firewalls. The needed ports must be open for communications to occur.

- **Number of expected users:** How many users will use the application within each cell (BSS)? All of the information gathered about throughput and utilization levels must be multiplied by the number of users to ensure that the cell can handle the load.

With this information gathered, you can make an effective recommendation for the number of APs that is required, the size of the cells and the number of users per AP. As you can see, to plan for an effective WLAN, many elements must be considered. The factors of throughput, utilization, latency tolerance, and the number of users have the most significant impact on your decisions for the number of APs and the number of users allowed in each cell and, therefore, the cell sizes.

> Control cell sizes by managing the output power of the APs. With reduced output power, the cell size or area in which clients can connect to the AP is constrained. Smaller cell sizes are often required with large numbers of users or data-intensive applications.

Chapter Summary

In this chapter, you learned about various client device types and the features and capabilities they offer. You also learned about important factors that must be considered in a client population on a network. These factors include single-band vs. multi-band support, supported PHYs, supported channels, security options and applications used. In the next and final chapter, everything you have learned will be brought together as you consider organizational requirements matched to WLAN features and functions.

Points to Remember

Remember the following important points:

- Most laptops have internal wireless NICs that are often upgradeable with mini-PCIe, half mini-PCIe or NGFF (M.2).

- Laptops and desktops can also use USB Wlan adapters.

- Using an internal PCI adapter on a desktop may not provide optimal wireless communications due to the location of the antennas.

- Tablets and mobile phones do not offer replaceable wireless adapters; therefore, the specifications must be known and accommodated.

- Many tablets and mobile phones are single-stream devices; this is intentional in order to reduce battery drainage.

- Most 2.4 GHz devices support all available channels in a regulatory domain or at least channels 1 through 11.

- 5 GHz devices often do not support the middle range of channels from 100 to 144.

- Channel 144 was first introduced in the 802.11ac amendment to the 802.11 standard.

- Channels 36 to 48 and 149 to 165 are supported by the vast majority of wireless clients.

- Multi-band devices support at least the 2.4 GHz and 5 GHz bands, and some newer devices also support the 60-GHz band, though they are few at the time of writing.
- Latency, throughput, utilization and the number of users are key factors in determining the needed number of APs and cell sizes in a WLAN deployment.

Review Questions

1. What kind of client devices should be purchased if at all possible?
   a. 802.11g
   b. 802.11n
   c. Multi-band
   d. 802.11b

2. What kind of client device typically has replaceable wireless NICs?
   a. Laptop
   b. Tablet
   c. Mobile phone
   d. Specialty device

3. What is the maximum number of spatial streams that are supported by most mobile phones?
   a. 1
   b. 2
   c. 3
   d. 4

4. Why is an internal wireless PCI NIC sometimes problematic for desktop computers?
   a. The antennas are often located under the user's desk
   b. They do not support modern PHYs
   c. They are USB based
   d. They are mini-PCIe based

5. Why should you update to the latest drivers for a USB device?
   a. To enable 802.11 operations
   b. To make it compatible with USB 3.0
   c. To enable all possible features
   d. To disable WEP

6. What PHY do most wireless door locks support as the highest data rate PHY?
    a. 802.11b
    b. 802.11a
    c. 802.11g
    d. 802.11ac

7. What three bands may be supported in modern multi-band devices?
    a. 100 MHz
    b. 2.4 GHz
    c. 5 GHz
    d. 60 GHz

8. What is required to accomplish the highest speeds that are advertised for an AP?
    a. Client devices supporting matching capabilities
    b. An RF channel with no RF activity at all
    c. Maximum output power on the AP
    d. Minimum output power on the AP

9. What channel is typically not supported by 802.11n and earlier PHYs?
    a. 40
    b. 100
    c. 144
    d. 165

10. When documenting applications, what metric is important to calculate the number of APs and cell sizes that are required?
    a. Ports used
    b. Antenna length
    c. Number of expected users
    d. Protocols used (TCP or UDP)

Review Answers

1. **C is correct.** Multi-band devices are the best choice. This allows for access to 5 GHz APs.

2. **A is correct.** Laptop computers use mini-PCIe, half mini-PCIe, and NGFF (M.2) adapters, which are usually replaceable.

3. **B is correct.** Two spatial streams are the maximum number supported by most mobile phones and tablets, and many still only support one spatial stream.

4. **A is correct.** Because the antennas protrude from the back of the desktop computer and the computer is often placed under the desk on the floor, the antennas are not optimally located.

5. **C is correct.** When purchasing USB adapters, it is not uncommon for the shipping driver to lack support for some features. For example, several multi-band devices have been sold over the years that require driver updates to enable 5 GHz operations.

6. **C is correct.** Sadly, most wireless 802.11 door locks only work with 2.4 GHz and support up to ERP (802.11g).

7. **B, C, and D are correct.** Today, multi-band devices support 2.4 GHz, 5 GHz, and 60 GHz.

8. **A is correct.** In order to achieve the highest speeds that an AP supports, the client devices must match features such as spatial streams and PHYs supported. Additionally, sufficient signal strength must be at the client location.

9. **C is correct.** Channel 144 was first introduced in 802.11ac and is usually not supported by 802.11n and older devices.

10. **C is correct.** Four factors are important when considering the number of APs and sizes for wireless cells: latency tolerance, throughput requirements, utilization and number of users.

Chapter 9 – WLAN Requirements

Objectives Covered

4.1 Understand issues in common vertical markets
- 4.1.1   Standard Enterprise Offices
- 4.1.2   Healthcare
- 4.1.3   Hospitality
- 4.1.4   Conference Centers
- 4.1.5   Education
- 4.1.6   Government
- 4.1.7   Retail
- 4.1.8   Industrial
- 4.1.9   Emergency Response
- 4.1.10   Temporary Deployments
- 4.1.11   Small Office/Home Office (SOHO)
- 4.1.12   Public Wi-Fi

4.2 Gather information about existing networks
- 4.2.1   Network diagrams
- 4.2.2   Wi-Fi implementations
- 4.2.3   Neighbor networks
- 4.2.4   Available network services
- 4.2.5   PoE availability

4.3 Discover coverage and capacity needs
- 4.3.1   Define coverage areas
- 4.3.2   Define capacity zones

4.5 Determine the need for outdoor coverage networks and bridge links
- 4.5.1   Bridge link distance and required throughput
- 4.5.2   Outdoor areas requiring coverage
- 4.5.3   Use cases for outdoor access

4.6 Define security constraints
- 4.6.1   Regulatory
- 4.6.2   Industry standards and guidelines

4.6.3 Organizational policies
4.7 Discover use cases and access types
    4.7.1 Authorized users
    4.7.2 Onboarded guest access
    4.7.3 Public Wi-Fi
4.8 Match organizational goals to WLAN features and functions

As you begin this final chapter, it is important to understand one key reality: The final knowledge domain of the CWS exam, *Understand Organizational Goals*, is dependent on the knowledge of the preceding eight chapters. That is, objectives 4.1 through 4.8 are not only covered in this chapter, but in the knowledge that has been acquired throughout the preceding chapters. The goal of this chapter is to review the knowledge gained from the previous chapters while adding extra insights as well. You will begin by exploring various vertical markets and then move on to specific requirements of networks. By the end of this chapter, you will have a basic understanding of the information that should be gathered in order to effectively match organizational goals to WLAN features and functions.

Common Vertical Markets

Wi-Fi is ubiquitous today in that it is used in every industry and government organization category. For this reason, it is important for the CWS to understand common vertical markets and their needs. At times, they overlap, and at times they vary in requirements. The following vertical markets should be considered by the CWS candidate:

- **Standard Enterprise Offices:** Office space is typically inclusive of carpeted flooring, office partitions, internal wall structures and often long hallways. Using properly located and configured APs with internal antennas will typically suffice. The device and application pool must be discovered to implement a proper design, but, in most cases, nothing more than the right number of internal antenna APs in the right locations with the right configurations will be required.

- **Healthcare:** Healthcare deployments are similar to office space deployments; coverage is required in most places, and standard office-type devices can be used; however, specialty medical equipment, push-to-talk devices, and facility design elements that provide special challenges are also present and introduce issues that are not seen in standard office deployments. The facility often includes special areas with walls that do not allow for typical RF propagation, particularly in X-ray rooms with metal-lined walls and in the many areas where elevators are used. These areas must be given special consideration for coverage needs. Additionally, many of the specialty devices may still use older PHY solutions, such as 802.11b, 802.11g, and 802.11a. Finally, healthcare regulations may demand special considerations for security and safety as well.

- **Hospitality:** The hospitality industry includes hotels and restaurants. While restaurants are typically deployed in a similar manner as standard office deployments, often with fewer APs, hotels have special considerations. The coverage must reach into the hotel rooms and also function in the hallways. APs must be placed to accommodate both locations. While some hotels have chosen to deploy APs in the hallways, this is not usually the best practice. In most cases, APs placed in some proper number of hotel rooms will cover the rooms and hallways as well. With wall-plate APs, they can often be installed in every room with low output power. Finally, special consideration must be given to access policies—for example, a captive portal may or may not be used, and authentication may or may not be required. Pre-shared key authentication (e.g., WPA2) seems like a good solution, but it results in the requirement of a shared passphrase for all guests, so the communications can be easily decrypted. In most cases, the hospitality WLAN is simply an open network, and guests should be informed to use encryption (with a VPN) if they will transmit sensitive data or to use HTTPS for communications. This can be communicated through the usage agreement screen that is displayed upon initial connection.

- **Conference Centers:** Conference centers should be considered medium- to high-density deployments. Since more clients will require WLAN access in a given square footage area, more APs are needed, and, at times, directional antennas may be required to accommodate the higher density of devices.

Additional challenges are introduced because the furniture and even walls are often changed on a day-to-day basis, and neighboring networks within the area often change as guests bring in their own networks.

> Problems in conference centers and arenas also include the density of people impacting RF propagation and the introduction of rogue devices into the space. A policy should be in place that requires authorization for the installation of new, temporary wireless networks.

- **Education:** Educational environments introduce new challenges. The walls are often thick to allow for sound dampening and simply because of older building structures. Classrooms may hold a high density of students, and they may have personal devices as well as school-issued devices. Some vendors recommend one AP per classroom for these reasons, and this is often required, but the CWS should understand the specific implementation and recommend what is needed. In some cases, a design that uses an AP in every classroom is appropriate, and in others (perhaps most), fewer APs will be required. It is also important to remember that schools have libraries and, due to the many books and other materials, RF propagation is often minimized in these spaces.

> A hidden node problem occurs when two STAs can hear the AP, but they cannot hear each other. When APs are placed in hallways, this problem increases due to the attenuation of typically thick school walls between the APs and clients.

- **Government:** Government installations often require security clearance for the installation technicians and may have special security requirements for the network implementation. For example, the security solution may have to comply with the Federal Information Processing Standard (FIPS) 140-2, which defines rules for cryptographic systems in all United States Federal agencies that deploy cryptographic systems. Given that 802.11 networks use encryption on the wireless links, such networks deployed in Federal agencies must comply with FIPS 140-2, which basically requires AES encryption be used on the network. Therefore, WPA (TKIP/RC4) is not allowed in many government deployments, and WPA2 (CCMP/AES) must be used. FIPS 14-2 simply states that an approved algorithm must be used. FIPS 197 defines AES as that approved algorithm, which replaced DES in 2001.

> *FIPS 140-2* defines four security levels. All four levels require an approved encryption algorithm (AES), and levels 2-4 add requirements, like tamper-resistant enclosures, identity-based authentication and so on. Figure 9.1 shows the detailed requirements of FIPS 140-2.

- **Retail:** Retail environments today often act as a public hotspot for customers and also have private Wi-Fi networks for inventory management, sales processing and more. The public hotspot will require all considerations documented later in this section regarding *Public Wi-Fi*. For the private network, device pools must be considered. If wireless barcode scanners and communications devices are used, you must determine the PHYs supported and other important characteristics like required signal strength and any real-time communications if push-to-talk is used. For payment processing, you may be required to comply with PCI-DSS (Payment Card Industry-Data Security Standard), which will demand separation of the payment processing network from the rest of the network and the use of authentication and encryption. Finally, depending on the type of retail shelving and racks, RF propagation may vary greatly from open office spaces, and thorough testing and validation should be performed by design and installation engineers.

- **Industrial:** Industrial installations include warehouses and manufacturing environments. Such installations include issues such as propagation model alterations due to shelving and the nature of the stocks, large metal machinery and frequently changing environments (for example, full shelves on Monday and nearly empty shelves on Thursday in some scenarios). The WLAN design must accommodate for these issues, and it is common to use APs with external antennas to propagate signals down hallways between shelving racks and/or to deploy more APs with lower output power even though the density of devices may be low. Each warehouse or manufacturing installation will be very unique, and such environments are typical candidates for pre-deployment full-site surveys as well as post-installation validation surveys. It is also important to remember that non-802.11 RF activity is common in these installations. Incidental energy from machinery and intentional radiation from video cameras and other non-802.11 devices will usually warrant a spectrum analysis scan.

	Security Level 1	Security Level 2	Security Level 3	Security Level 4
Cryptographic Module Specification	Specification of cryptographic module, cryptographic boundary, Approved algorithms, and Approved modes of operation. Description of cryptographic module, including all hardware, software, and firmware components. Statement of module security policy.			
Cryptographic Module Ports and Interfaces	Required and optional interfaces. Specification of all interfaces and of all input and output data paths.		Data ports for unprotected critical security parameters logically or physically separated from other data ports.	
Roles, Services, and Authentication	Logical separation of required and optional roles and services.	Role-based or identity-based operator authentication.	Identity-based operator authentication.	
Finite State Model	Specification of finite state model. Required states and optional states. State transition diagram and specification of state transitions.			
Physical Security	Production grade equipment.	Locks or tamper evidence.	Tamper detection and response for covers and doors.	Tamper detection and response envelope. EFP or EFT.
Operational Environment	Single operator. Executable code. Approved integrity technique.	Referenced PPs evaluated at EAL2 with specified discretionary access control mechanisms and auditing.	Referenced PPs plus trusted path evaluated at EAL3 plus security policy modeling.	Referenced PPs plus trusted path evaluated at EAL4.
Cryptographic Key Management	Key management mechanisms: random number and key generation, key establishment, key distribution, key entry/output, key storage, and key zeroization.			
	Secret and private keys established using manual methods may be entered or output in plaintext form.		Secret and private keys established using manual methods shall be entered or output encrypted or with split knowledge procedures.	
EMI/EMC	47 CFR FCC Part 15. Subpart B, Class A (Business use). Applicable FCC requirements (for radio).		47 CFR FCC Part 15. Subpart B, Class B (Home use).	
Self-Tests	Power-up tests: cryptographic algorithm tests, software/firmware integrity tests, critical functions tests. Conditional tests.			
Design Assurance	Configuration management (CM). Secure installation and generation. Design and policy correspondence. Guidance documents.	CM system. Secure distribution. Functional specification.	High-level language implementation.	Formal model. Detailed explanations (informal proofs). Preconditions and postconditions.
Mitigation of Other Attacks	Specification of mitigation of attacks for which no testable requirements are currently available.			

Figure 9.1: Table 1 from the FIPS 140-2 Standard (Detailed knowledge of this table is not required for the CWS exam.)

Variations occurring throughout the day in industrial deployments must be considered. For example, it is common to have large volumes of material placed in loading dock areas that are temporarily awaiting shipment. These materials can have an adverse effect on the network. These areas may also have dramatic density fluctuations since barcode scanners are used to receive large shipments.

- **Emergency Response:** In most cases, emergency response systems use reserved frequencies and non-802.11 equipment. For example, in the United States, the band from 4.940 to 4.990 GHz is used for public safety and requires a license to operate. However, during disaster recovery situations, it is also possible for temporary 802.11-based networks to be deployed. These networks may use backhaul links that take advantage of the public safety band or some other wireless technology. The local network will be 802.11-based, and the links back to central offices will be licensed bands.

Licensed bands require that the installing organization acquire a license to operate wireless equipment in the band. 802.11 networks use unlicensed bands and can be installed by anyone; they only require that the equipment operates within regulatory constraints.

- **Temporary Deployments:** Construction locations, outdoor events, and other locations often require temporary deployments. A site survey of such locations should be performed to identify other networks in the area and select the best channels for installation. Additionally, the backhaul link must be considered. While emergency response installations may be able to use the public safety band, other deployments will likely use LTE cellular backhauls to gain Internet connectivity or to redirect traffic to a central location. Wired backhaul connections are seldom used for temporary deployments.

- **Small Office/Home Office (SOHO):** A SOHO deployment typically consists of a single AP or, more often, wireless routers. They are often consumer-grade devices from Linksys, Belkin, NETGEAR, D-Link and other vendors. It is becoming more common to see vendors like Ubiquiti, EnGenius, and TP-Link in these spaces as well. Additionally, vendors such as Cisco, Aerohive, Aruba Networks, Mojo Networks, and Ruckus have provided SOHO-class devices for use in these installations. Most SOHO installations use PSK security and rarely implement a RADIUS server for this reason, though some third-party RADIUS server providers (accessed across the Internet) are sometimes used. It is important that SOHO deployments be implemented with proper channel selection and output power settings. While channel 6 is a very commonly selected channel, it is rarely the best channel in the 2.4-GHz band because of the fact that it is a common default setting. Instead, the installer should locate the best available channels among 1, 6 and 11 and configure the AP/wireless router to operate on that channel. It is also common to use the wireless AP/router provided by the Internet Service Provider (ISP). These devices may be managed by the ISP, but they usually allow for local reconfiguration according to the customer's needs.

- **Public Wi-Fi:** Public Wi-Fi deployments must provide coverage where required; while this is easily understood, they must also ensure that they provide the needed capacity. If the network performance is poor or access is too complex, the customers will choose to go elsewhere. Since one of the primary purposes of such public Wi-Fi networks is often to draw in customers, it could be tempting to use an open authentication method. This would simply allow users onto the network with no authentication. However, for marketing purposes or legal concerns, many establishments choose to use other authentication methods, such as social logins (e.g., logging in with a Facebook account). Once the social connection is established, the customer is granted access but may also receive information about sales and other promotional materials. Captive portals are often used in public hotspot deployments for these purposes. They also enforce an agreement with the use policy.

> Users of public hotspots often use a Virtual Private Network (VPN) solution to protect their communications. The VPN encrypts all traffic and forwards it to the VPN server, which then forwards it to the destination.

Gathering Essential Information

The largest task in matching organizational goals to WLAN features and functions is defining the condition of the environment and the needs of the organization. Without this essential first step, it is impossible to recommend or select a solution that will work for the organization. As a CWS, you do not have to define every detailed technical requirement, but you should have a general understanding of the organization's needs and be able to match them to the solutions that you have learned about in this book. In this section, the essential information that must be gathered is addressed.

Existing Networks

You must consider several factors related to existing networks. These include existing Wi-Fi implementations, neighbor networks, available network service, PoE availability and network diagrams. This section will explain all of these and their importance.

Wi-Fi Implementations

If the target facility or organization has an existing Wi-Fi implementation, this means that the new install is either an upgrade, a replacement (sometimes called a forklift upgrade), or an expansion of that network. The CWS should know what existing networks are in place as well as the reason for the upgrade or expansion. In all cases, the following should be discovered:

- Bands used in the existing networks.

- Bands required for the upgrade or expansion.

- Areas covered by the existing WLAN and areas to be covered by the new or expanded WLAN.

- Client population currently in use.

- New clients expected to be deployed in the future.

- Current WLAN management solution (e.g., controller, cloud, autonomous).

- A site survey should be performed to ensure that the expansion or upgrade results in the required coverage, capacity, and functionality demanded by the deployment.

> Remember, a *site survey* includes discovering existing RF signals in the space, channels already occupied, building structures and more in order to provide a better design for the WLAN upgrade, replacement or expansion.

Neighbor Networks

Neighbor networks are networks that surround the location of a planned WLAN upgrade, replacement or expansion. These networks should be identified in the site survey. The following neighbor network information should be acquired:

- Bands and channels used at specific locations.
- The signal strength of the wireless activity in the detected locations.
- Any non-Wi-Fi RF activity that is impacting the target location.

With this information, you can effectively plan the WLAN implementation. For example, if the neighbor network is using channels 1, 6 and 11, which are detectable at a particular location, and channels 1 and 11 are detected at -62 dBm while channel 6 is detected at -88 dBm, you will be forced to use channel 6 in that area if 2.4-GHz coverage is required. Alternatively, you can only use 5-GHz coverage in that area if all client devices will support 5-GHz channels.

> In addition to neighbor network channel signal strength, the WLAN engineer may factor in the utilization of the channel. For example, if channel 11 is at -72 with 18% utilization and channels 6 and 1 are at -83 dBm with 60% utilization, channel 11 is likely the better choice.

Available Network Services

Several network services are required to allow WLANs to function properly. These services include the following:

- **DHCP:** The Dynamic Host Configuration Protocol (DHCP) is used to configure the IP settings for nodes on the network. It will provide an IP address, subnet mask, default gateway and DNS server IP address among other possible settings. It is important to ensure that the DHCP pool (the collection of IP addresses it can distribute) is large enough to accommodate the number of clients that the WLAN will introduce. DHCP pool depletion (having an insufficient number of available IP addresses for the client population) is a common problem for WLANs.

- **DNS:** The Domain Name System (DNS) is used to map hostnames (e.g., computer01.company.com) to IP addresses. It is common to attempt a connection to a network device based on the host name and fail because the name mapping does not exist in the DNS server. DNS is also used by many APs to locate the appropriate controller.

- **Time Servers:** Time servers (also called Network Time Protocol [NTP] servers) are required by many network services, including authentication services. If RADIUS is used as an authentication server and an LDAP user database (such as Active Directory) is used, an available Time Server with proper configuration is required.

- **Switching:** APs connect to network switches for access to the wired network. The switching must be properly configured including setting the VLAN for the connected devices, enabling or disabling port security, ensuring proper operating mode (access mode and trunk mode) and enabling PoE when required. Additionally, switch speeds must meet the demands of the wireless network. For example, modern APs should connect to 1-Gbps switch ports at a minimum.

- **Routing:** Routers interconnect IP networks. For example, if one network is identified as 192.168.10.0 and another is identified as 192.168.11.0, a router will be required to allow traffic to flow between the two networks. The routing must be configured to allow the appropriate traffic through the routing interfaces.

These basic services will be required in most WLAN implementations. The only listed services that may not be required in smaller SOHO implementations without the use of WPA2-Enterprise is the time server and routing. All other services are required, even in a SOHO implementation, though the switching may be integrated into a wireless router.

PoE Availability

Power over Ethernet (PoE) is used to power most APs that are deployed in enterprise installations today. In fact, many APs come with no power cord (traditional wall output power cords) and either ship with a PoE injector or assume PoE will be provided in most installations. When deploying with PoE, the following must be considered:

- The PoE standard required by the APs. While PoE is simply part of the 802.3 Ethernet standard, two iterations were introduced in 802.3af and 802.3at. While an 802.3af-compliant device requires only 12.95 watts of power or less, an 802.3at-compliant device may require up to 25.5 watts of power.

- Cable runs for PoE-powered devices are usually shorter than the cable required for wall-outlet-powered devices because the PoE power can create interference for the data signals traveling on the same Ethernet cable.

- The power budget of the switch must be factored into the deployment. If a switch has a PoE budget of 150 watts, this indicates that a limited number of switch ports can be enabled for PoE. More switch ports can be enabled for 802.3af than for 802.3at. If full power is required, an 802.3af port will output 15.4 watts of power, and an 802.3at port will output 30 watts of power. Remember that the output power at the Power Source Equipment (PSE) is greater than the received power at the Powered Device (PD).

- Midspan PoE injectors are often used when the switch does not support PoE or does not have PoE ports or power budget available. Figure 9.2 shows a midspan PoE injector.

- When an AP is connected to a known PoE switch and power is not provided, check the settings for the switch port to ensure that PoE is enabled and verify that the connected switch port is a PoE port. Most PoE switches offer a limited number of ports on which PoE operates.

Figure 9.2: Midspan PoE Injector (EDIMAX GP-101IT 802.3at Gigabit Injector)

An additional device called a PoE splitter has become popular in recent years. It is used to power a non-PoE device using a PoE cable to deliver power to the location. Figure 9.3 (courtesy of EDIMAX) shows a PoE injector used to power a PoE device (a Wi-Fi AP) and a PoE injector used with a PoE splitter to power a non-PoE device. Notice that the splitter breaks the power away from the data at the point of the splitter by providing DC output power from the splitter to the non-PoE device. The image shows the EDIMAX GP-101IT as the PoE midspan injector and the EDIMAX GP-101SF as the PoE splitter.

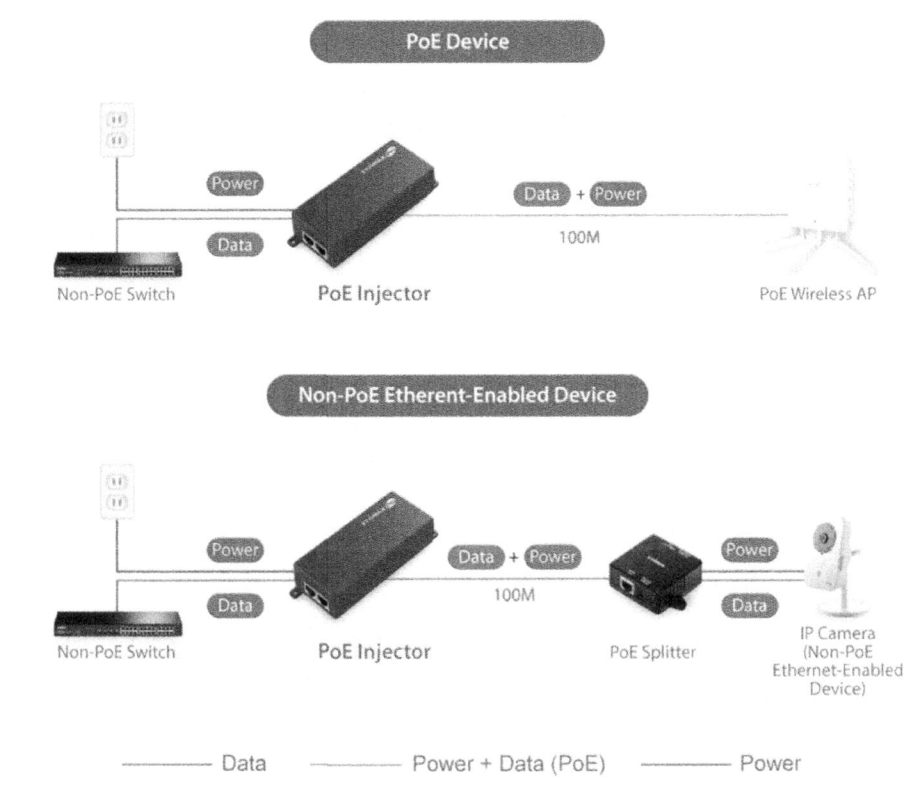

Figure 9.3: PoE Injectors used for both PoE and non-PoE Devices (Courtesy of EDIMAX)

> The maximum length for a PoE injector-based cable (whether from a switch [endspan] or injector [midspan] device) is approximately 100 meters or 328 feet. Sometimes you will need to shorten the cable length depending on the quality of the Ethernet cable, so targeting a maximum of 300 feet is best.

Network Diagrams

The final component in this section is network diagrams. Network diagrams illustrate and indicate the structure of the network. When overlaid with floor plans, they show specific locations of devices. Diagrams with floor plans are often best since they help you determine the length of cable from switches to APs, but they are not always available. However, if you have a logical diagram (one that does not map to the floor plan) and a separate floor plan, you can gather additional information about device locations and identify the needed information yourself.

Figure 9.4 shows an example logical network diagram. It is valuable in that it indicates existing services available on the network such as LDAP directory services (the AD (Active Directory server)), DNS availability, DHCP availability, switches, and routers.

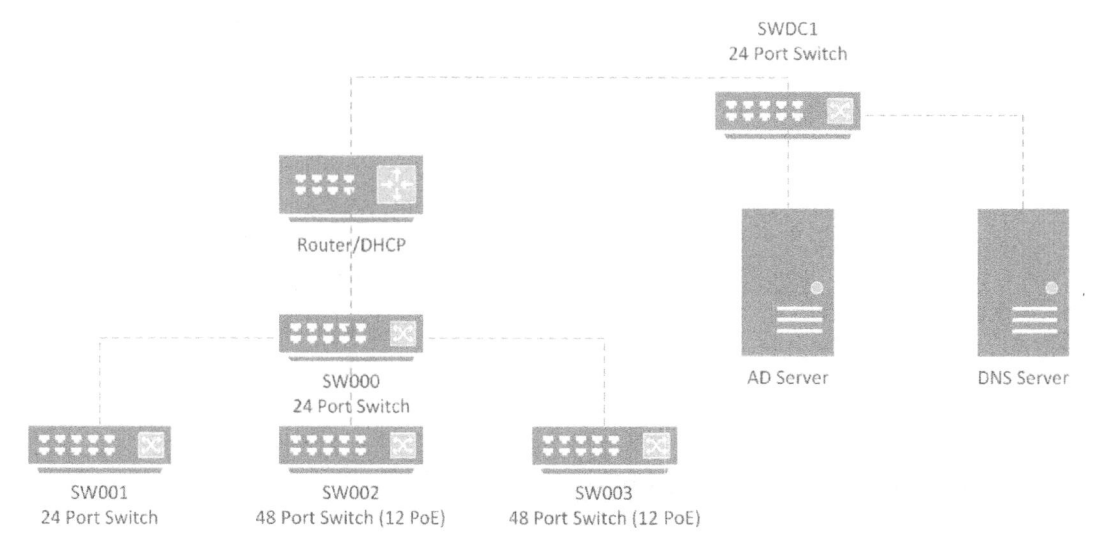

Figure 9.4: Logical Network Diagram

Figure 9.5 shows the same logical network overlaid on a floor plan. This can be performed quickly if you have the Visio (or another diagramming tool) document for the logical network. Simply drag the items to the proper physical location, and now you have a better idea of the needed cable lengths and available switch and PoE ports.

> Network diagrams are most useful when they reveal the locations for existing APs and provide information to use in the estimation of cable lengths for new APs. Without this information, they are primarily used to understand services that are available on the network.

Coverage and Capacity Needs

Coverage is easy; capacity is hard. This statement is important to remember and emphasize with decision makers. They must understand that anyone can accomplish covering an area with an RF signal, but covering it with the right signal and doing it while providing enough APs to achieve capacity and quality takes expertise. For this reason, the CWS should always recommend or use a trained wireless systems engineer (SE) in the WLAN design process.

Imagine you are asked to enhance an existing WLAN to improve capacity. You will have to consider the channels already in use, the number of concurrent connections required in the capacity areas and the distance required between new APs and existing APs. At times, existing APs will have to be relocated to allow for proper deployment. Additionally, existing APs may have to be replaced. Enhancing an existing WLAN can be a challenging endeavor, and in many cases, it is more complex than implementing an entirely new WLAN.

Coverage areas should be clearly defined. They should be defined based on the required signal strength, and the required signal strength may not be the same for all coverage areas. For example, some areas may require VoIP connections, and others may not. The areas requiring VoIP connections will demand a -65 dBm to -70 dBm coverage specification, depending on vendor requirements. The other areas may tolerate a -70 dBm to -75 dBm specification, depending on the applications used and other needs.

Defining general coverage areas is no longer sufficient for designing WLANs. Capacity zones or areas must also be defined. When defining capacity zones, the number of devices and the demanded throughput per device should be identified. With this information, the proper number of APs can be selected.

When designing for capacity, the maximum transmit power and range are not the key factors. The most important factor is to select APs that support the newest PHYs so that client devices supporting them can get on and off the medium as quickly as possible.

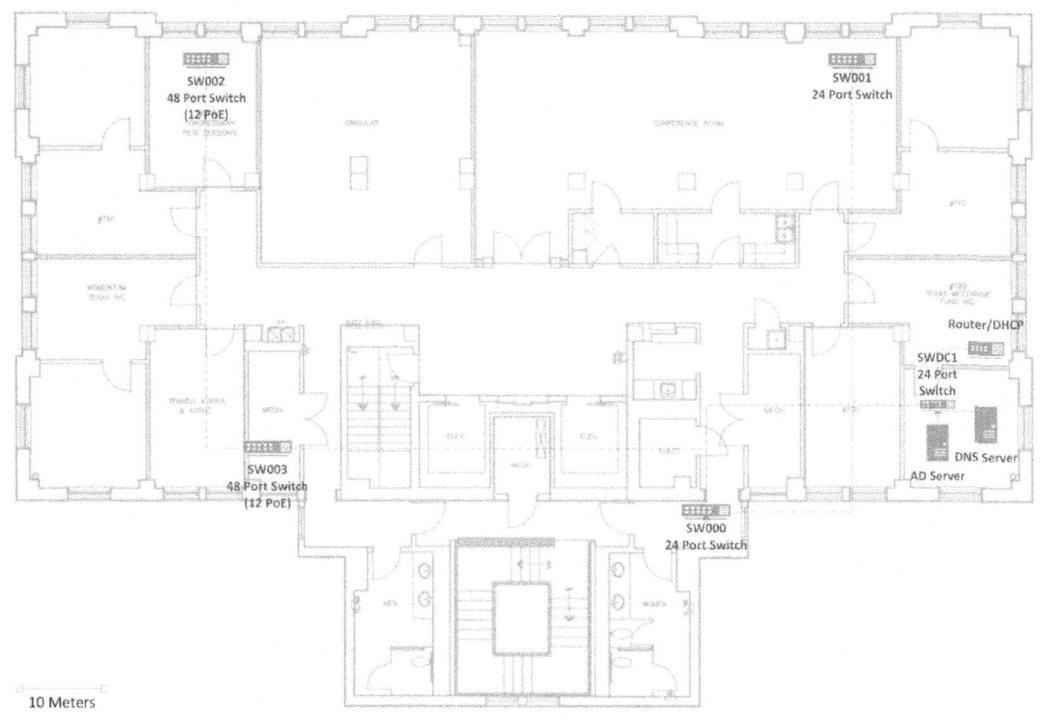

Figure 9.5: Floor Plan Mapped to the Logical Diagram

Figure 9.6 shows a floor plan with coverage and capacity areas defined. In this example, coverage areas are defined based on signal strength requirements. Additionally, desired coverage areas are indicated so that the budget can be managed and fewer APs can be deployed to remain within budget if that is required. By not covering some areas, costs can be reduced. However, most organizations find that such desired-only areas are actually required coverage areas when the users begin to use the network. That is, the users complain of lacking coverage in those areas. The CWS should encourage the decision maker to deploy wireless coverage anywhere an employee may be performing work-related tasks.

Note in Figure 9.6 that the capacity zones have been defined with information related to the number of devices and the throughput required. This information will assist the designer in the design process. For example, some APs have specifications that limit the number of connected clients, and many engineers know that the true limit is much lower for a given AP if exceptional performance is desired.

Using information like that provided in the floor plan diagram shown in Figure 9.6 assists the engineer in designing a well-performing and scalable WLAN. Scalability is an important factor because new use cases always arise after the WLAN is installed. To accommodate these use cases, the implemented WLAN should have room for growth as new clients are added to the network.

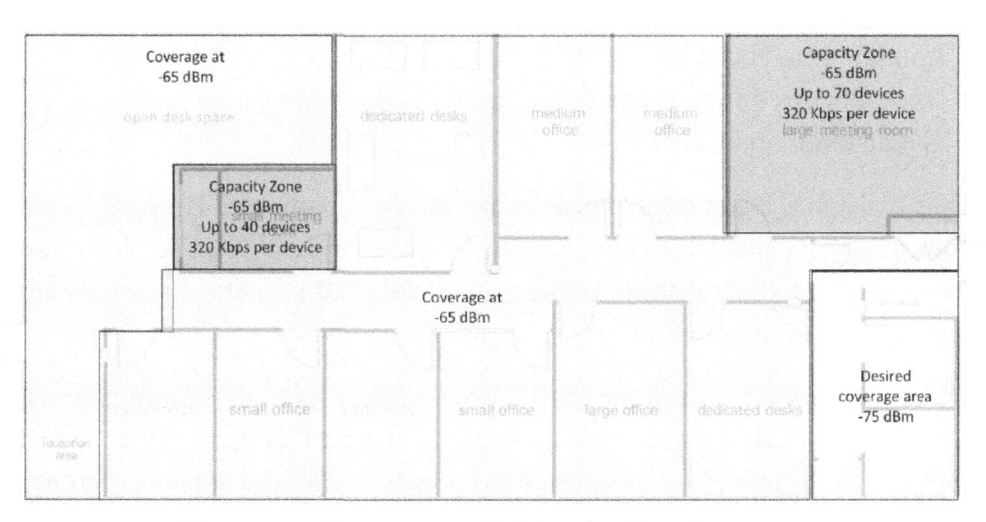

Figure 9.6: Coverage and Capacity Requirements

It may also be important to define coverage requirements for each network type. For example, guest networks are not usually required in areas such as the accounting department, network operations center and other sensitive areas.

Outdoor Coverage Networks and Bridge Links

Outdoor network links include coverage for clients as well as bridge links. Outdoor use cases include (at least) the following:

- Coverage between buildings for roaming users.
- Coverage in outdoor eating and lounge areas.
- Outdoor coverage for public hotspots.

The coverage in outdoor spaces is easy to implement, but capacity can be more challenging because of the distance that RF signals travel without walls and other barriers to limit them. Given the higher number of channels in the 5-GHz band, it is easier to deploy high-capacity WLANs outdoors in that band.

Wireless bridge links are used to connect networks across the wireless link. To create a properly functioning bridge link, several factors must be properly planned:

- The output power of the AP/bridge.
- Antenna gain and type.
- Losses incurred from cables from the AP/bridge to the antenna.
- The distance of the link.

The link distance is important because of free space path loss. RF energy spreads as it travels, and the available energy at any given location is lessened as the receiver gets farther from the transmitter. The general rule is that as the distance doubles, the signal halves in power.

Regulatory constraints limit the output power of the radio in most regulatory domains as well as the Equivalent Isotropically Radiated Power (EIRP), which is the power in the intended direction of propagation after the antenna gain. Using higher gain antennas allows for longer bridge links.

It is essential that bridge links have RF line of sight, which is not the same as the visual line of sight—that is, RF line of sight requires more clearance due to the spreading of the RF wave. It is for this reason that bridge antennas are often placed on the top of buildings or even on poles or towers.

Finally, it is best to avoid allowing clients to connect to the APs at each end of the bridge link. The remote bridge will not be able to hear client transmissions at the other end of the link, and significant interference will occur if clients are allowed to connect. Each AP/bridge should be dedicated to the bridge link or links in which it participates.

Building proper bridge links is a technical process that involves the calculation of link budgets and determination of Fresnel zones. A qualified engineer should always be involved in bridge link design and implementation.

Security Constraints

All WLANs will be limited by security constraints. A security constraint controls the methods that can be used for WLAN authentication and may disallow WLAN installations in some areas of a facility, depending on organizational policies. Three types of security constraints should be considered: regulatory, industry and organizational.

Regulatory constraints are those imposed by a local regulatory agency. For example, the federal government in the United States requires the protection of patient health information in healthcare organizations under penalty of law. The regulation that imposes these constraints is called the Health Insurance Portability and Accountability Act (HIPAA). It was first implemented to guarantee employees that their insurance could be transferred with them when they moved to different companies. Later, it was amended to require protection of patient health information. Therefore, a WLAN implemented in a healthcare environment that transmits health information on the WLAN should use encryption to protect this data.

Industry standards and guidelines are not enforced by the local government, but they may be enforced through other means. For example, the Payment Card Industry-Data Security Standard (PCI-DSS) is not enforced by law, but credit card companies may not allow you to process their cards if you do not comply. PCI-DSS requires that WLANs involved in payment card processing be protected with proper encryption and separated from other networks by a firewall.

The final security constraint is organizational policies. These internal policies are developed to ensure compliance with regulatory and industry standards and also to diminish risk. While regulatory and industry constraints protect other organizations and individuals, organizational policies go further to protect the organization that is implementing the WLAN from additional risks and therefore costs.

Security policies should always be evaluated before changing settings in WLAN equipment. All devices should be configured according to these policies since they provide protection from risk, regulatory non-compliance, and industry non-compliance if they are properly designed.

Use Cases and Access Types

The final element to consider is the various use cases for the WLAN and access types. A use case is a defined purpose for the use of the WLAN. For example, a wireless client device can be used to scan barcodes on boxes that come in or that are removed from inventory. Such a barcode scanner is a common use case for WLAN clients. When exploring use cases, the throughput requirements and utilization should be defined. Throughput will be defined in Kbps or Mbps, and utilization will be defined as a percentage per client. With this information and the number of clients in a cell implementing the use case, you can determine the capacity needs of those clients.

It is important to remember that many client devices implement multiple use cases. For example, a mobile VoIP handset may sometimes act as a VoIP phone, sometimes as a messaging client and sometimes as an email client. All use cases must be considered.

The following are common use cases with some estimates (which will vary depending on parameters of the use case) of common throughput and utilization metrics:

- VoIP: 100-200 Kbps, 10% utilization
- E-Mail: 50-100 Kbps, 10% utilization
- Messaging: 5-20 Kbps, 5% utilization
- Data transfer: 1-5 Mbps, 10-20% utilization
- Database access: 100-500 Kbps, 10-15% utilization
- Web browsing: 100-5000 Kbps, 15-25% utilization

In addition to the use cases, the access type should be considered in order to implement the proper security solutions. Three access types are considered in the CWS exam: authorized users, onboarded guest access, and public Wi-Fi.

Authorized users should be granted access in compliance with organizational policies. Today, these policies should require WPA2-PSK or WPA2-Enterprise with rare exceptions. This requirement will demand CCMP/AES for encryption. When PSK is used, the passphrase should be changed at some defined interval. When Enterprise is used, the infrastructure must make an authentication server available, such as a RADIUS server.

Onboarded guest access differs from what we often consider public Wi-Fi in that it requires user registration. This registration may involve a simple captive portal system that requires a user to log on with a pre-staged account, or users may have to register and request account approval before being granted access. With either method, it is likely that some administrative staff will be involved with the approval processes. To avoid this, some organizations implement social authentications that gather information from Facebook or other social media accounts and instantly grant access.

Finally, public Wi-Fi will usually be implemented as an open network with no security at the Wi-Fi layers. Hotspot 2.0 certified devices may become more available in the future, but they are not the most commonly used public hotspot solutions today. Some hotspots, often too many of them, will implement a captive portal to either gather information from the user or require their agreement to the acceptable use policy of the organization.

> *Passpoint* (release 2) is a Wi-Fi Alliance certification that allows automatic use authentication when connecting to a partnering Hotspot 2.0 site. The authentication occurs seamlessly if the user information is already provisioned on the mobile device and the link is encrypted.

A key factor in public Wi-Fi deployments is performance. If you deploy a public hotspot, you should ensure that it will perform well for the expected number of clients or not deploy it at all. A poorly performing Wi-Fi hotspot is bad for the brand and image of the organization. In fact, it is often better to simply not provide Wi-Fi than to provide poor Wi-Fi.

Chapter Summary

In this chapter, you explored the many issues that must be considered when matching organizational requirements to WLAN features and functions. With this information, which aggregated much of the knowledge acquired in preceding chapters, you are well prepared to recommend or select appropriate WLAN devices that will meet the needs of target organizations. At this point, you have explored all areas of the CWS objectives and should have sufficient knowledge to pass the CWS exam. If you desire more information, consider the e-learning available for CWS candidates at CWNP.com. Additionally, you can use the practice exams, which are also available on our website, to test your readiness for exam day. Congratulations, and good luck on your exam!

Points to Remember

Remember the following important points:

- Existing Wi-Fi implementations impact new installations because they generate RF signals that must be accounted for when planning for expansion or upgrade.

- Neighbor WLANs must also be considered based on the RF signals they radiate.

- Several network services are required for the typical WLAN to function properly, including DHCP, DNS, time servers, switching, and routing.

- PoE requirements must be met, including PoE type, cable length constraints and the power budget in switches.

- Coverage should be defined by area including signal strength requirements.

- Capacity zones should be defined including signal strength requirements, the number of client devices and the throughput demands of those devices.

- Network diagrams are useful when identifying available network services and, potentially, required cable lengths.

- Bridge links must be carefully planned and include factors such as required received signal strength, antenna gain, link distance and output power.

- A key outdoor use case is roaming coverage for users that move between buildings.

- Outdoor capacity can be more challenging since walls, doors and other items do not exist to attenuate the signals.

- Security constraints in WLANs include regulatory, industry and organizational policy constraints.

- Changes made to WLAN configuration settings should be validated against policies.

- Public Wi-Fi networks may use captive portals for user authentication or simply to require user agreement with acceptable use policies.

Review Questions

1. What facility issue may impact RF behavior in hospitals around X-ray rooms?
    a. RF interference from the X-ray machine
    b. Metal-lined walls
    c. Glass doors
    d. RF-resistant paint used on the flooring

2. What factor impacts capacity in a conference center or arena when large events are held there that may not impact it when smaller events are held?
    a. Wall materials
    b. Carpet materials
    c. Density of people
    d. Use of 900 MHz phones

3. What should always be performed in a WLAN deployment even if it is just an expansion of an existing WLAN?
    a. Site survey
    b. AP replacement

    c. Client replacement

    d. Analysis of activity in the 60 GHz band

4. What is a common problem related to DHCP when deploying WLANs?

    a. Improper switch port configurations

    b. Missing hostname mappings

    c. Pool depletion

    d. Non-Wi-Fi interference

5. What kind of device provides power across a PoE cable to a non-PoE device?

    a. Midspan injector

    b. Endspan injector

    c. PoE splitter

    d. Non-PoE switch

6. Why would a PoE cable be limited to a shorter length than a typical Ethernet data-only cable?

    a. Power waves do not travel as far as electromagnetic waves

    b. The PoE power can cause interference on the Ethernet data wires

    c. PoE causes refraction of the RF waves

    d. PoE causes diffraction of the RF waves

7. What is often missing from a logical network diagram?

    a. Physical locations of devices

    b. Services available on the network

    c. PoE port availability on switches

    d. DHCP availability on the LAN

8. Which one of the following is not a factor considered when selecting an AP for capacity-based deployments?

    a. Throughput capability

    b. Number of supported client devices

    c. RF band of operation

    d. Maximum output power settings

9. Why are outdoor networks often more challenging from a capacity perspective than indoor networks?
    a. The signal does not travel as far
    b. Outdoor networks cannot use the 5-GHz band
    c. Outdoor networks lack environmental components that attenuate the signal
    d. APs designed for outdoor deployment support fewer client devices

10. What use case requires QoS and 100-200 Kbps of throughput with low latency?
    a. E-Mail
    b. VoIP
    c. Database access
    d. Web browsing

Review Answers

1. **B is correct.** X-ray rooms have metal-lined walls (or other materials) that prohibit RF communications and must be accounted for in hospital deployments.

2. **C is correct.** Humans absorb RF energy, and when more people are in a space, it will impact the signal strength and propagation, though a significant increase in people is required to have a negative impact.

3. **A is correct.** A site survey should always be performed to gather important information like existing channels used, non-Wi-Fi interferers and RF propagation behaviors in the targeted coverage areas.

4. **C is correct.** DHCP pools offer a limited number of IP addresses. When WLANs are added to existing networks, it is not uncommon for the pools to be depleted, resulting in an apparent connection problem for client devices that is actually a network problem, while the wireless link is functioning properly.

5. **C is correct.** A PoE splitter will split the power from a PoE source back into separate power and data cables and can be used to power a non-PoE device (e.g., a network camera).

6. **B is correct.** PoE cables sometimes require shorter cables than typical Ethernet cables because the PoE power can cause interference with the Ethernet data wires in the cable.

7. **A is correct.** In many cases, logical network diagrams provide no information about physical locations. This information is required to determine necessary cable lengths for the WLAN deployment.

8. **D is correct.** Because capacity-based deployments rarely use output power levels on APs that are close to the maximum available, the maximum output power available is not usually a factor.

9. **C is correct.** In outdoor deployments, walls, windows, and doors are not available to attenuate signals, which makes the capacity design more challenging.

10. **B is correct.** VoIP, based on information in this chapter and preceding chapters, requires QoS, 100-200 Kbps of throughput and low latency to perform acceptably.

Glossary: A CWNP Universal Glossary

40 MHz Intolerant: A bit potentially set in the 802.11 frame allowing STAs to indicate that 40 MHz channels should not be used in their BSS or in surrounding networks. The bit is processed only in the 2.4 GHz band.

4-Way Handshake: The process used to generate encryption keys for unicast frames (Pairwise Transient Key (PTK)) and transmit encryption keys for group (broadcast, multicast) (Group Temporal Key (GTK)) frames using material from the 802.1X/EAP authentication or the pre-shared key (PSK). The PTK and GTK are derived from the Pairwise Master Key (PMK) and Group Master Key (GMK) respectively.

802.11: A standard maintained by the IEEE for implementing and communicating with wireless local area networks (WLANs). Regularly amended, the standard continues to evolve to meet new demands. Several Physical Layer (PHY) methods are specified and the Medium Access Control (MAC) sublayer is also specified.

802.11a: An 802.11 amendment that operates in the 5GHz band. It uses OFDM modulation and is called the OFDM PHY. It can support data rates of up to 54 Mbps.

802.11aa: An 802.11 amendment that added support for robust audio and video streaming through MAC enhancements. It specifies a new category of station called a Stream Classification Service (SCS) station. The SCS implementation is optional for a WMM QoS station.

802.11ac: An 802.11 amendment that operates in the 5GHz band. It uses MU-MIMO, beamforming, and 256 QAM technology, up to 8 spatial streams and OFDM modulation. Support is included for data rates up to 6933.3 Mbps.

802.11ae: An 802.11 amendment that provides prioritization of management frames. It defines a new Quality of Service Management Frame (QMF). When the QMF service is used, some management frames may be transmitted using an access category other than the one used for voice (AC_VO). When communicating with stations that do not support the QMF service, the station uses access category

AC_VO to transmit management frames. When QMF is supported, the beacon frame includes a QMF Policy element.

802.11ah: An 802.11 draft that specifies operations in the sub-1 GHz range. Frequencies used vary by regulatory domain. The draft supports 1, 2, 4, 8 and 16 MHz channels with OFDM modulation.

802.11ax: An 802.11 draft that will support bi-directional MU-MIMO, higher modulation rates and sub-channelization. It is too early to know the final details of this amendment at the time of writing; however, it is planned to operate in the 2.4 GHz and 5 GHz band.

802.11b: An IEEE 802.11 amendment that operates in the 2.4GHz ISM band. It uses HR/DSSS and earlier technology. It can support data rates of up to 11Mbps.

802.11e: An 802.11 amendment, now incorporated into the most recent roll-up, that provided quality of service extensions to the wireless link through probabilistic prioritization based on the contention window. The Wi-Fi Multimedia (WMM) certification is based on this amendment.

802.11g: An IEEE 802.11 amendment that operates in the 2.4GHz ISM band. It uses ERP-OFDM and earlier technology. It can support data rates of up to 54Mbps.

802.11i: An 802.11 amendment, now incorporated into the most recent roll-up, which provided security enhancements to the standard and resolved weaknesses in the original WEP encryption solution. It provided for TKIP/RC4 (now deprecated) and CCMP/AES cipher suites and encryption algorithms.

802.11n: An IEEE 802.11 amendment that operates in the 2.4 ISM and 5GHz UNII/ISM bands. It uses MIMO, HT-OFDM and earlier technology. It can support data rates of up to 600Mbps.

802.11k: An IEEE 802.11 amendment that specifies and defines WLAN characteristics and mechanisms.

802.11r: An IEEE 802.11 amendment that enables roaming between access points.

802.11u: An IEEE 802.11 amendment that adds features for mobile communication devices such as phones and tablets.

802.11w: An IEEE 802.11 amendment to increase security for the management frames.

802.11y: An IEEE 802.11 amendment that allows registered stations to operate at a higher power output in the 3650-3700 MHz band.

802.1X: 802.1X is an IEEE standard that uses the Extensible Authentication Protocol (EAP) framework to authenticate devices attempting to connect to the LAN or WLAN. The process involves the use of a supplicant to be authenticated, authenticator, and authentication server.

802.11 State Machine: The 802.11 state machine defines the condition of the connection of a client STA to another STA and can be in one of three states: Unauthenticated/Unassociated, Authenticated/Unassociated, or Authenticated/Associated.

802.3: A set of standards maintained by the IEEE for implementing and communicating with wired Ethernet networks and including Power over Ethernet (PoE) specifications.

AAA Framework: Authentication, Authorization, and Accounting is a framework for monitoring usage, enforcing policies, controlling access to computer resources, and providing the correct billing amount for services.

AAA Server Credential: The AAA server credential is the validation materials used for the server. When mutual authentication is required, a server certificate is typically used as the AAA server credential.

Absorption: Occurs when an obstacle absorbs some or all of a radio wave's energy.

Access Category (AC): An access category is a priority class. 802.11 specifies four different priority classes – voice (AC_VO), video (AC_VI), best effort (AC_BE), and background (AC_BK).

Access Layer Forwarding: Data forwarding that occurs at the access layer, also called *distributed data forwarding*. The data is distributed from the access layer directly to the destination without passing through a centralized controller.

Access Point: An access point (AP) is a device containing a radio that is used to create an access network, bridge network or mesh network. The AP contains the Distribution System Service.

Access Port: An AP used for mesh networks and that connects to the wired or wireless network at the edge of the mesh.

Acknowledgement Frame: A frame sent by the receiving 802.11 station confirming the received data.

Access Control List (ACL): ACLs are lists that inform a STA or user what permissions are available to access files and other resources. ACLs are also used in routers and switches to control packets allowed through to other networks.

Active Mode: A power-save mode in which the station never turns the radio off.

Active Scanning: A scanning (network location) method in which the client broadcasts probe requests and records the probe responses in order to determine the network with which it will establish an association.

Active Survey: A wireless survey conducted on location that involves measuring throughput rates, round trip time, and packet loss by connecting devices to an AP and transmitting data during the survey.

Ad-Hoc Mode: The colloquial name for an Independent Basic Service Set (IBSS). STAs connect directly with each other, and an AP is not used.

Adjacent Overlapping Channels: Adjacent overlapping channels are channels whose bands interfere with their neighboring channels on the primary carrier frequencies. Non-overlapping channels are channels whose bands do not interfere with neighboring channels on the primary carrier frequencies.

Adjacent Channel Interference (ACI): ACI occurs when channels near each other (in the frequency domain) interfere with one another due to either partial frequency overlap on primary carrier frequencies or excessive output power.

AES (Advanced Encryption Standard): The encryption cipher used with CCMP and WPA2 providing improved security over WEP/RC4 or TKIP/RC4.

AID: Association ID (AID) is an identification assigned by a wireless STA (AP) to another STA (client) in order to transmit the correct data to that device in an Infrastructure Basic Service Set.

AirTime Fairness: Transmits more frames to client STAs with higher data rates than those with lower data rates so that the STAs get fair access to the air (medium) instead of having to wait for slower data rate STAs.

Aggregated MAC Protocol Data Units (A-MPDU): A-MPDU transmissions are created by transmitting multiple MPDUs as one PHY frame as opposed to A-MSDU transmissions, which are created by passing multiple MSDUs down to the PHY layer as a single MPDU.

Aggregated MAC Service Data Unit (A-MSDU): See *Aggregated MAC Protocol Data Unit*.

Amplification: The process of increasing a signal's power level.

Amplifier: A device intended to increase the power level of a signal.

Amplitude: The power level of a signal.

Antenna: A device that converts electric power into radio waves and radio waves into electric power.

Association: The condition wherein a client STA is linked with an AP for frame transmission through the AP to the network.

Announcement Traffic Indication Message (ATIM): A traffic indication map (sent in a management frame) in an Ad-Hoc (IBSS) network to notify other clients of pending data transfers for power saving purposes.

Attenuation: The loss of signal strength as an RF wave passes through a medium.

Attenuator: A device that intentionally reduces the strength of an RF signal.

Authentication: The process of the user or device identity validation.

Authentication and Key Management (AKM): The protocols used to authenticate a client STA on a WLAN and generate an encryption key for use in frame encryption.

Authentication Server: The authentication server validates the client before allowing access to the network. In an 802.1X/EAP implementation for WLANs, the authentication server is often a RADIUS server.

Authenticator: The device that provides access to authentication services in order to allow connected devices to access network resources. In an 802.1X/EAP implementation for WLANs, the authenticator is typically the AP or controller.

Automatic Power Save Delivery (APSD): APSD is a power saving method which uses both scheduled (S-APSD) and unscheduled (U-APSD) frame delivery methods. S-APSD sends frames to a power save STA from the AP at a planned time. U-APSD sends frames to a power save STA from the AP when the STA sends a frame to the AP. The frame from the STA is considered a trigger frame.

Autonomous AP: An AP that can perform security functions, RF management, and configuration without the need for a centralized WLAN controller or any other control platform.

Azimuth Chart: A chart showing the radiation pattern of an antenna as viewed from the top of the antenna. Also called an H-Plane Chart or H-Chart.

Backoff timer: The timer used during CSMA/CA to wait for access to the medium, which is selected from the contention window.

Band Steering: A method used by vendors to encourage STAs to connect to the 5 GHz band instead of the 2.4 GHz band, which is more congested. Typically implemented by ignoring probe requests for some period of time before allowing connection to the 2.4 GHz radio by clients known to have a 5 GHz radio based on previous connections to the AP or controller.

Bandwidth: The frequencies used for transmission of data. For example, a 20 MHz wide channel has 20 MHz of bandwidth.

Basic Service Area (BSA): The coverage area provided by an AP wherein client STAs may connect to the AP to transmit data on the WLAN or through the AP to the network.

Basic Service Set (BSS): An AP and its associated STAs. Identified by the BSSID.

Basic Service Set Identification (BSSID): The ID for the BSS. Often the MAC address of the AP STA. When multiple SSIDs are used, another MAC address-like BSSID is generated.

Beacon Frame: A frame transmitted periodically from an AP that indicates the presence of a BSS network and contains capabilities and requirements of the BSS. Also colloquially called a beacon instead of the full phrase, beacon frame.

Beamforming: Directing radio waves to a specific area or device by manipulating the RF waveforms within the different radio chains.

Beamwidth: The width of the radiated signal lobe from the antenna in the intended direction of propagation. It is usually measured at the point where 3 dB of loss is experienced.

Bill of materials (BOM): A list of the materials and licenses required to assemble a system, in the case of WLANs, including APs, controllers, PoE injectors, licenses, etc.

Bit: A basic unit of information for computer systems. A bit can have a value of 1 or 0. Used in binary math.

Block Acknowledgement: An acknowledgment frame that groups together multiple ACKs instead of transmitting each individual ACK when a block transmission has been received.

Bridge: A device used to connect two networks. Wireless bridges create the connection across the wireless medium.

BSS Transition: Roaming that occurs between two BSSs that are part of the same ESS.

Byte: A basic unit of information that typically consists of 8 bits. Also called an octet.

Capacity: The number of clients and applications a network or AP can handle.

Captive Portal: Authentication technique that re-routes a user to a special webpage to verify their credentials before allowing access to the network. Commonly used in hotel and guest networks.

Guest Networks: A segregated network that is designed for use by temporary visitors.

CardBus: A PCMCIA PC Card standard interface that supports 32-bits and operates at speeds of up to 33 MHz. It is primarily used in laptops.

Carrier Frequencies: The frequency of a carrier signal or the frequencies used to modulate information.

Carrier Sense Multiple Access (CSMA): CSMA is a protocol that allows a node to detect the presence of traffic before sending data on a shared network. Used in CSMA/CA.

Carrier Sense Multiple Access with Collision Avoidance (CSMA/CA): CSMA/CA is the method in 802.11 networks in which a node only sends data if the shared network is idle in order to avoid collisions.

CCMP: Counter Cipher Mode with Block Chaining Message Authentication Code Protocol (CCMP) is a key management solution that provides for improved security over WEP.

CCMP/AES: CCMP used with AES, as it is in 802.11 networks, is a key management and encryption protocol that provides more security than WEP. It is based on the AES standard and uses a 128-bit key and 128-bit block size.

Centralized Forwarding: Every forwarding decision is made by a centralized forwarding engine, such as the WLAN controller.

Certificate Authority (CA): A server that validates the authenticity of a certificate used in authentication and encryption systems. The CA may issues certificates, or it may authorize other servers to do the same.

CompactFlash (CF): Originally produced in 1994 by SanDisk, CF is a flash memory mass storage device format that can support up to 256 GB. CF devices can also function as 802.11 WLAN adapters.

Channel: A specified range of frequencies used in the 802.11 standard used by devices to communicate on the network. Channels are commonly 20, 40, 80 and 160 MHz in width in WLANs. Newer standards will support 1, 2, 4, 8 and 16 MHz channels in sub-1 GHz networks.

Channel Width: The range of frequencies a single channel encompasses.

Clear Channel Assessment (CCA): CCA is a feature defined in the IEEE 802.11 standard that allows a client to determine the idle or busy state of the medium based on energy levels of a frame or raw energy levels as specified in each PHY.

Client Utilities: Software installed on devices that allows the device to connect to, authenticate with and participate in a WLAN.

Co-Channel Interference (CCI): Congestion caused by the normal operations of CSMA/CA when multiple BSSs exist on the same channel. Commonly called co-channel congestion (CCC) today as well.

Collision Avoidance (CA): A method in which devices attempt to avoid simultaneous data transmissions in order to prevent frame collisions. Used in CSMA/CA.

Coding: A process used to encode bits to be transmitted on the wireless medium such that error recovery can be achieved. Part of forward error correction (FEC) and defined in the modulation and coding schemes (MCSs) from 802.11n forward.

Containment: A process used against a detected rogue AP to prevent any connected clients from accessing the network.

Contention Window: A number range defined in the 802.11 standard and varying by QoS category from which a number is selected at random for the backoff timer in the CSMA/CA process.

Control Frame: An 802.11 frame that is used to control the communications process on the wireless medium. Control frames include RTS frames, CTS frames, PS-Poll frames, and ACK frames.

Controlled Port: In an 802.1X authentication system, the virtual port that allows all frames through to the network, but only after authentication is completed.

Controller-Based AP: An AP managed by a centralized controller device. Also called a lightweight AP or thin AP.

Coverage: 1) The colloquial term used for the BSA of an AP. 2) The requirement of available WLAN connectivity throughout a facility, campus or area. Often specified in minimum signal strength as dBm; for example, -67 dBm.

Clear-to-Send (CTS) Frame: A CTS frame sent from one STA to another to indicate that the other STA can transmit on the medium. The duration value in the CTS frame is used to silence all other STAs by setting their NAV timers.

Data Frame: An 802.11 frame specified for use in carrying data based on the general frame format. Also used for some signaling purposes as null data frames.

Data Rate: The rate at which data is sent across the wireless medium. Typically represented as megabits per second (Mbps) or gigabits per second (Gbps). The data rate should not be confused with throughput rate, which is a measurement of Layer 4 throughput or useful user data.

dBd (decibel to dipole): A relative measurement of antenna gain compared to a dipole antenna. Calculated as 2.14 dB greater than dBi as a dipole antenna already has 2.14 dBi gain.

dBi (decibel to isotropic): A relative measurement of antenna gain compared to a theoretical isotropic radiator. When necessary, calculated as 2.14 dB less than dBd.

dBm (decibel to milliwatt): An absolute measurement of the power of an RF signal based on the definition of 0 dBm = 1 milliwatt (mW).

Distributed Coordination Function (DCF): A protocol defined in 802.11 that uses carrier sensing, backoff timers, interframe spaces and frame duration values to diminish collisions on the wireless medium.

Elevation Chart: A chart showing the radiation pattern of an antenna as viewed from the side antenna. Also called an E-Plane Chart or E-Chart.

Deauthentication Frame: A notification frame sent from an 802.11 STA to another STA in order to terminate a connection between them.

Decibel (dB): A logarithmic, relative unit used when measuring antenna gain, signal attenuation, and signal-to-noise ratios. Strictly defined as 1/10 of a bel.

Delay: The time it takes for a bit of data to travel from one node to another. Also called latency.

Delivery Traffic Indication Message (DTIM): A message sent from an AP to clients in the Beacon frame indicating that it has data to transmit to the clients specified by the AIDs.

Differentiated Services Code Point (DSCP): A Layer 3 QoS marking system. IP packets can include DSCP markings in the headers. Eight precedence levels, 0-7, are defined.

Diffraction: The bending of waves around a very large object in relation to the wave.

Direct-Sequence Spread Spectrum (DSSS): A modulation technique where data is coupled with coding that spreads the data across a wide frequency range. Provides 1 or 2 Mbps data rates in 802.11 networks.

Disassociation Frame: A frame sent from one STA to another in order to terminate the association.

Distributed Forwarding: See *Access Layer Forwarding*. Also called, *distributed data forwarding*.

Distribution System (DS): The system that connects a set of BSSs and LANs such that an ESS is possible.

Distribution System Medium (DSM): The medium used to interconnect APs through the DS such that they can communicate with each other for ESS operations using either wired or wireless for the DS connection.

Domain Name System (DNS): A protocol and service that provides hostname resolution (looking up the IP address of a given hostname) and recursive IP address lookups (finding the hostname of a known IP address). Also, colloquially used to reference the server that provides DNS lookups.

Driver: Software that allows a computer to interact with a hardware device such as a WLAN adapter.

Duty Cycle: A measure of the time a radio is transmitting or a channel is consumed by a transmitting device.

Dynamic Frequency Selection (DFS): A setting on radios that dynamically changes the channel selection based on detected interference from radar systems. Many 5 GHz channels require DFS operations.

Dynamic Rate Switching (DRS): The process of reducing a client's data rate as frame transmission failures occur or signal strength decreases. DRS results in lower data rates but fewer transmissions required to successfully transmit a frame.

Encryption: The process of converting data into a form that unauthorized users cannot understand by encoding the data with an algorithm and a key or keys.

Enhanced Distributed Channel Access (EDCA): An enhancement to DCF introduced in 802.11e that implements priority based queuing for transmissions in 802.11 networks based on access categories.

Equivalent Isotropically Radiated Power (EIRP): The output power required of an isotropic radiator to equal the measured power output from an antenna in the intended direction of propagation.

Extended Rate Physical (ERP): A physical layer technology introduced in 802.11g that uses OFDM (from 802.11a) in the 2.4 GHz band and offers data rates up to 54 Mbps.

Extended Service Set (ESS): A group of one or more BSSs that are interconnected by a DS.

Extensible Authentication Protocol (EAP): An authentication framework that defines message formats for authentication exchanges used by 802.1X WLAN authentication solutions.

Fade Margin: An amount of signal strength, in dB, added to a link budget to ensure proper operations.

Fast Fourier Transform (FFT): A mathematical algorithm that takes in a waveform as represented in the time or space domain and shows it in the frequency domain. Used in spectrum analyzers to show real-time views in the frequency domain (Real-time FFT).

Fragmentation: The process of fragmenting 802.11 frames based on the fragmentation threshold configured. Fragmented frames have a greater likelihood of successful delivery in the presence of sporadic interference.

Frame Aggregation: A feature in the IEEE 802.11n PHY and later PHYs that increases throughput by sending more than one frame in a single transmission. Aggregated MSDUs or aggregated MPDUs may be supported.

Frame: A well-defined, meaningful set of bits used to communicate management and control information on a network or transfer payloads from higher layers. Frames are defined at the MAC and PHY layer.

Free Space Path Loss: The natural loss of amplitude that occurs in an RF signal as it propagates through space and the wavefront spreads.

Fresnel Zones: Ellipsoid shaped zones around the visual LoS in a wireless link. The first Fresnel zone should be 60% clear and would preferably be 80% clear to allow for environmental changes.

Frequency: The speed at which a waveform cycles in a second.

Full Duplex: A communication system that allows an endpoint to send data to the network at the same time as it receives data from the network.

Gain: The increase in signal strength in a particular direction. Can be accomplished passively by directing energy into a smaller area or actively by increasing the strength of the broadcasted signal before it is sent to the antenna.

Group Key Handshake: Used to transfer the GTK among STAs in an 802.11 network if the GTK requires updating. Initiated by the AP/controller in a BSS.

Group Master Key (GMK): Used to generate the GTK for encryption of broadcast and multicast frames and is unique to each BSS.

Group Temporal Key (GTK): Used to encryption broadcast and multicast frames and is unique to each BSS.

Guard Interval (GI): A period of time between symbols within a frame used to avoid intersymbol interference.

Half Duplex: A communication system that allows only sending or receiving data by an endpoint at any given time.

Hidden Node: The problem that arises when nodes cannot receive each other's frames, which can lead to packet collisions and retransmissions.

High Density: A phrase referencing a WLAN network type that is characterized by large numbers of devices requiring access.

Highly-Directional Antenna: An antenna, such as a parabolic dish or grid antenna, that has a high gain in a specified direction and a low beamwidth measurement as compared to semi-directional and omnidirectional antennas.

High Rate Direct Sequence Spread Spectrum (HR/DSSS): An amendment-based PHY (802.11b) that increase the data rate in 2.4 GHz from the original 1 or 2 Mbps to 5.5 and 11 Mbps while maintaining backward compatibility with 1 and 2 Mbps.

High Throughput (HT): An amendment-based PHY (802.11n) that increased the data rate up to 600 Mbps and added support for transmit beamforming and MIMO.

Hotspot: A term referencing a wireless network connection point that is typically open to the public or to paid subscribers.

Independent Basic Service Set (IBSS): A set of 802.11 devices operating in ad-hoc (peer-to-peer) mode without the use of an AP.

Institute of Electrical and Electronics Engineers (IEEE): A standardization organization that develops standards for multiple industries including the networking industry with standards such as 802.3, 802.11, and 802.16.

Intentional Radiator: Any device that is purposefully sending radio waves. The signal strength of the intentional radiator is measured at the point where energy enters the radiating antennas.

Interference: In WLANs, an RF signal or incidental RF energy that is radiated in the same frequencies as the WLAN and that has sufficient amplitude and duty cycle to prevent 802.11 frames from successful delivery.

Interframe Space (IFS): A time interval that must exist between frames. Varying lengths are used in 802.11 and are referenced as DIFS, SIFS, EIFS and AIFS in common use.

Internet Engineering Task Force (IETF): An open group of volunteers develops Internetworking standards through a request for comments (RFC) documents. Examples include RADIUS, EAP, and DNS.

Isotropic Radiator: A theoretical antenna that spreads the radiation equally in every direction as a sphere. None exist in reality, but the concept is used to measure relative antenna gain in dBi.

Jitter: The variance in delay between packets sent on a network. Excessive jitter can result in poor quality for real-time applications such as voice and video.

Jumbo Frame: An Ethernet frame that contains more than 1500 bytes of payload and up to 9000 to 9216 bytes.

Latency: The time taken for data to move between places. Typically synonymous with the delay in computer networking.

Layer 1: The physical layer (PHY) that is responsible for framing and transmitting bits on the medium. In 802.3 and 802.11 the entirety of Layer 1 is defined.

Layer 2: The data-link layer that deals with data frames moving within a local area network (LAN). In 802.3 and 802.11, the MAC sublayer of Layer 2 is defined.

Layer 3: The network layer where packets of data are routed between sender and receiver. Most modern networks use Internet Protocol (IP) at Layer 3.

Layer 4: The transport layer where segmentation occurs for upper layer data and TCP (connection-oriented) and UDP (connectionless) are the most commonly used protocols.

Lightning Arrestor: A device that can redirect ambient energy from a lightning strike away from the attached equipment.

Line of sight (LoS): When existing, the visual path between two ends. RF LoS is different from visual LoS. RF LoS does not require the same clear path for the remote receiver to hear the signal. When creating bridge links, visual LoS is often the starting point.

Link Budget: The measurement of gains and losses through an intentional radiator, antenna and over a transmission medium.

Loss: The reduction in the amplitude of a signal.

MAC filtering: A common setting that only allows specific MAC addresses onto a network. Ineffective against knowledgeable attackers because the MAC address can be spoofed to impersonate authorized devices.

Management Frame: A frame type defined in the 802.11 standard that encompasses frames used to manage access to the network including beacon, probe request, prober response, authentication, association, reassociation, deauthentication, and disassociation frames.

Master Session Key (MSK): A key derived between an EAP client and EAP server and exported by the EAP method. Used to derive the PMK, which is used to derive the PTK. The MSK is used in 802.1X/EAP authentication implementations. In personal authentication implementations, the PMK is derived from the pre-shared key.

Maximal Ratio Combining (MRC): A method of increasing the signal-to-noise ratio (SNR) by combining signals received on multiple radio chains (multiple antennas and radios).

Mesh: A network that uses interconnecting devices to form a redundant set of connections offering multiple paths through the network. 802.11s defined mesh for 802.11 networks.

Mesh BSS: A basic service set that forms a self-contained network of mesh stations.

Milliwatt (mW): A unit of electrical energy used in measuring the output power of RF signals in WLANs. A mW is equal to 1/1000 of a watt (W).

Mobile User: A user that physically moves while connected to the network. The opposite of a stationary user.

Modulation: The process of changing a wave by changing its amplitude, frequency, and/or phase such that the changes represent data bits.

Modulation and Coding Scheme (MCS): Term used to describe the combination of the radio modulation scheme and the coding scheme used when transmitting data, first introduced in 802.11n.

MPDU: A MAC protocol data unit (MPDU) is a portion of data to be delivered to a MAC layer peer on a network, and it is data prepared for the PHY layer by the MAC sublayer. The MAC sublayer receives the MSDU from upper layers on transmission and creates the MPDU. It receives the MPDU from the lower layer on receiving instantiation and removes the MAC header and footer to create the MSDU for the upper layers.

MSDU: A MAC service data unit is a portion of transmitted data to be handled by the MAC sublayer that has yet to be encapsulated into a MAC Layer frame.

Maximum Transmission Unit (MTU): The largest amount of data that can be sent at a particular layer of the OSI model. Typically set at layer 4 for TCP.

Multi-User MIMO (MU-MIMO): An enhancement to MIMO that allows the AP STA to transmit to multiple client STAs simultaneously.

Multipath: The phenomenon that occurs when multiple copies of the same signal reach a receiver based on RF behaviors in the environment.

Multiple Channel Architecture (MCA): A wireless network design using multiple channels strategically designed so that the implemented BSSs have minimal interference with one another.

Multiple Input/Multiple Output (MIMO): A technology used to spread a stream of data bits across multiple radio chains using spatial multiplexing at the transmitter and to recombine these streams at the receiver.

Narrowband Interference: Interference that covers a very narrow band of frequencies and typically not the full width of an 802.11 channel when used in reference to WLAN interferers.

Near-Far: A problem that occurs when a high-powered device is closer to the AP in a BSS, and a low powered device is farther from the AP. Most near-far problems are addressed with standard CSMA/CA operations in 802.11 networks.

Network Allocation Vector (NAV): The NAV is a virtual carrier sense mechanism used in CSMA/CA to avoid collisions and is a timer set based on the duration values in frames transmitted on the medium.

Network Segmentation: The process used to separate a larger network into smaller networks often utilizing Layer 3 routers or multilayer switches.

Noise: RF energy in the environment that is not part of the intentional signal of your WLAN.

Noise Floor: The amount of noise that is consistently present in the environment, which is typically measured in dBm.

Network Time Protocol (NTP): A protocol used to synchronize clocks in devices using centralized time servers.

Octet: A group of eight ones and zeros. An 8-bit byte. Sometimes simply called a byte.

Orthogonal Frequency Division Multiplexing (OFDM): A modulation technique and a named physical layer in 802.11 that provides data rates up to 54 Mbps and operates in the 5 GHz band. The modulation is used in all bands, but the named PHY operates only in the 5 GHz band.

Omni-Directional Antenna: An antenna that propagates in all directions horizontally. Creates a coverage area similar to a donut shape (toroidal). Also known as a dipole antenna.

Dipole Antenna: An antenna that propagates in all directions horizontally. Creates a coverage area similar to a donut (toroidal) shape. Also known as an omnidirectional antenna.

Open System Authentication: A simple frame exchange, providing no real authentication, used to move through the state machine in relation to the connection between two 802.11 STAs.

Opportunistic Key Caching (OKC): A roaming solution for WLANs wherein the keys derived from the 802.1X/EAP authentication are cached on the AP or controller such that only the 4-way handshake is required at the time of roaming.

OSI (Open Systems Interconnection) Model: A theoretical model for communication systems that works by separating the communications process into seven, well-defined layers. The seven layers are Application, Presentation, Session, Transport, Network, Data Link and Physical.

Packet: Data as represented at the network layer (Layer 4) for TCP communications.

Passive Gain: An increase in strength of a signal by focusing the signal's energy rather than increasing the actual energy available, such as with an amplifier.

Passive scanning: A scanning (network location) method wherein a STA waits to receive beacon frames from an AP which contain information about the WLAN.

Passive survey: A survey conducted on location that gathers information about RF interference, signal strength, and coverage areas by monitoring RF activity without active communications.

Passphrase Authentication: A type of access control that uses a phrase as the passkey. Also called personal in WPA and WPA2.

Phase: A measurement of the variance in arrival state between two copies of a waveform. Waves are said to be in phase or out of phase to some degree. The phase can be manipulated for modulation.

PHY: A shorthand notation for the physical layer which is the physical means of communication on a network to transmit bits.

Physical (PHY) Layer: The physical (PHY) layer refers to the physical means by which a message is communicated. Layer one of the OSI model.

PLCP: Physical Layer Convergence Protocol (PLCP) is the name of the service within the PHY that receives data from the upper layers and sends data to the upper layers. It is the interaction point with the MAC sublayer.

PMD: Physical Medium Dependent (PMD) is the service within the PHY responsible for sending and receiving bits on the RF medium.

PMK Caching: Stores the PMK, so a device only has to perform the 4-way handshake when connecting to an AP to which it has already connected.

Pairwise Master Key (PMK): The key derived from the MSK, which is generated during 802.1X/EAP authentication. Used to derive the PTK. Used in unidirectional communications with a single peer.

PoE Injector: Any device that adds Power over Ethernet (PoE) to ethernet cables. Come in two variants, endpoint (such as switches) and midspan (such as inline injectors).

Point-to-Multipoint (PtMP): A connection between a single point and multiple other points for wireless bridging or WLAN access.

Point-to-Point (PtP): A connection between two points often used to connect two networks via bridging.

Polarization: The technical term used to reference the orientation of antennas related to the electric field in the electromagnetic wave.

Power over Ethernet (PoE): A method of providing power to certain hardware devices that can be powered across the Ethernet cables. Specified in 802.3 as a standard. Various classes are defined based on power requirements.

PPDU: PLCP Protocol Data Unit (PPDU) is the prepared bits for transmission on the wired or wireless medium. Sometimes also called a PHY Layer frame.

Preauthentication: Authenticating with an AP to which the STA does not intend to immediately connect, so that roaming delays are reduced.

Pre-shared Key (PSK): Refers to any security protocol that uses a password or passphrase or string as the key from which encryption materials are derived.

Primary Channel: When implementing channels wider than 20 MHz in 802.11n and 802.11ac, the 20 MHz channel on which management and control frames are sent and the channel used by STAs not supporting the wider channel.

Probe Request: A type of frame sent when a client device wants information about APs in the area or is seeking a specific SSID to which it desires to connect.

Probe Response: A type frame sent in response to a probe request that contains information about the AP and the requirements of BSSs it provides.

Protected Management Frame (PMF): Frames used for managing a wireless network that are protected from spoofing using encryption. Protocol defined in the 802.11w amendment.

Protocol Analyzer: Hardware or software used to capture and analyze networking communications. WLAN protocol analyzers have the ability to capture 802.11 frames from the RF medium and decode them for display and analysis.

Protocol Decodes: The way information in captured packets or frames is interpreted for display and analysis.

PSDU: PLCP Service Data Unit (PSDU) is the name for the contents that are contained within the PPDU, the PLCP Protocol Data Unit. It is the same as the MPDU as perceived and received by the PHY.

PTK (Pairwise Transient Key): A key derived from the 4-way handshake and used for encryption only between two specific endpoints, such as an AP and a single client.

Quality of Service (QoS): Traffic prioritization and other techniques used to improve the end-user experience. IEEE 802.11e includes QoS protocols for wireless networks based on access categories.

QoS BSS: A BSS supporting 802.11e QoS features.

Radio Chains: A reference to the radio and antenna used together to transmit in a given frequency range. Multi-stream devices have multiple radio chains as one radio chain is required for each stream.

Radio Frequency (RF): The electromagnetic wave frequency range used in WLANs and many other wireless communication systems.

Radio Resource Management (RRM): Automatic management of various RF characteristics like channel selection and output power. Known by different terms among the many WLAN vendors, but referencing the same basic capabilities.

RADIUS: Remote Authentication Dial-In User Service (RADIUS) refers to a network protocol that handles AAA management which allows for authentication, authorization, and accounting (auditing). Used in 802.11 WLANs as the authentication server in an 802.1X/EAP implementation.

RC4 (Rivest Cipher 4): An encryption cipher used in WEP and with TKIP. A stream cipher.

Real-Time Location Service (RTLS): A function provided by many WLAN infrastructure and overlay solutions allowing for device location based on triangulation and other algorithms.

Reassociation: The process used to associate with another AP in the same ESS. May also be used when a STA desires to reconnect to an AP to which it was formerly connected.

Received Channel Power Indicator (RCPI): Introduced in 802.11k, a power measurement calculated as INT((dBm + 110) * 2). Expected accuracy is +/- 5 dB. Ranges from 0-220 are available with 0 equaling or less than -110 dBm and 220 equaling or greater than 0 dBm. The value is calculated as an average of all received chains during the reception of the data portion of the transmission. All PHYs support RCPI and, though 802.11ac does not explicitly list its formulation, it references the 802.11n specification for calculation procedures.

Received Signal Strength Indicator (RSSI): A relative measure of signal strength for a wireless network. The method to measure RSSI is not standardized though it is constrained to a limited number of values in the 802.11 standard. Many use the term RSSI to reference dBm, and the 802.11 standard uses terms like DataFrameRSSI and BeaconRSSI and defines them as the signal strength in dBm of the specified frames, so the common vernacular is understandable. However, according to the standard, "absolute accuracy of the RSSI reading is not specified" (802.11-2012, Clause 14.3.3.3).

Reflection: An RF behavior that occurs when a wave meets a reflective obstacle large than the wavelength similar to light waves in a mirror.

Refraction: An RF behavior that occurs as an RF wave passes through material causing a bending of the wave and possible redirection of the wavefront.

Regulatory Domain: A reference to geographic regions management by organizations like the FCC and ETSI that determine the allowed frequencies, output power levels and systems to be used in RF communications.

Remote AP: An AP designed to be implemented at a remote location and managed across a WAN link using special protocols.

Resolution Bandwidth (RBW): The smallest frequency that can be extracted from a received signal by a spectrum analyzer or the configuration of that frequency. Many spectrum analyzers allow for the adjustment of the RBW within the supported range of the analyzer.

Retry: That which occurs when a frame fails to be delivered successfully. A bit set in the frame to specify that it is a repeated attempt at delivery.

Return Loss: A measure of how much power is lost in delivery from a transmission line to an antenna.

RF Cables: A cable, typically coaxial, that allows for the transmission of electromagnetic waves between a transceiver and an antenna.

RF Calculator: A software application used to perform calculations related to RF signal strength values.

RF Connector: A component used to connect RF cables, antennas and transmitters. RF connectors come in many standardized forms and should match in type and resistance.

RF Coverage: Synonymous with coverage in WLAN vernacular. Reference to the BSA provided by an AP.

RF Link: An established connection between two radios.

RF Line of Sight (LoS): The existence of a path, possibly including reflections, refractions, and pass-through of materials, between two RF transceivers.

RF Propagation: The process by which RF waves move throughout an area including reflection, refraction, scattering, diffraction, absorption and free space path loss.

RF Signal Splitter: An RF component that splits the RF signal with a single input and multiple outputs. Historically used with some antenna arrays, but less common today in WLAN implementations.

RF Site Survey: The process of physically measuring the RF signals within an area to determine resulting RF behavior and signal strength. Often performed as a validation procedure after implementation based on a predictive model.

Roaming: That which occurs when a wireless STA moves from one AP to another either because of end-user mobility or changes in the RF coverage.

Robust Security Network (RSN): A network that supports CCMP/AES or WPA2 and optionally TKIP/RC4 or WPA. To be an RSN, the network must support only RSN Associations (RSNAs), which are only those associations that use the 4-way handshake. WEP is not supported in an RSN.

Robust Security Network Association (RSNA): An association between a client STA and an AP that was established through authentication resulting in a 4-way handshake to derive unicast keys and transfer group keys. WEP is not supported in an RSNA.

Rogue Access Point: An access point that is connected to a network without permission from a network administrator or other official.

Rogue Containment: Procedures used to prevent clients from associating with a rogue AP or to prevent the rogue AP from communicating with the wired network.

Rogue Detection: Procedures used to identify rogue devices. May include simple identification of unclassified APs or algorithmic processes that identify likely rogues.

Role-Based Access Control (RBAC): An authorization system that assigns permissions and rights based on user roles. Similar to group management of authorization policies.

RSN Information Element: A portion of the beacon frame that specifies the security used on the WLAN.

Request to Send/Clear to Send (RTS/CTS): A frame exchange used to clear the channel before transmitting a frame in order to assist in the reduction of collisions on the medium. Also used as a backward compatible protection mechanism.

RTS Threshold: The minimum size of a frame required to use RTS/CTS exchanges before transmission of the frame.

S-APSD: See *Automatic Power Save Delivery*.

Scattering: An RF behavior that occurs when an RF wave encounters reflective obstacles that are smaller than the wavelength. The result is multiple reflections or scattering of the wavefront.

Secondary Channel: When implementing channels wider than 20 MHz in 802.11n and 802.11ac, the second channel used to form a 40 MHz channel for data frame transmissions to and from supporting client STAs.

Semi-Directional Antenna: An antenna such as a Yagi or a patch that has a propagation pattern which maximizes gain in a given direction rather than an omnidirectional pattern, having a larger beamwidth than highly directional antennas.

Service Set Identifier (SSID): The BSS and ESS name used to identify WLAN. Conventionally made to be readable by humans. Maximum of 32 bytes long.

Signal Strength: A measure of the amount of RF energy being received by a radio. Often specified as the RSSI, but referenced in dBm, which is not the proper definition of RSSI from the 802.11 standard.

Single Channel Architecture (SCA): A WLAN architecture that places all APs on the same channel and uses a centralized controller to determine when each AP can transmit a frame. No control of client transmissions to the network is provided.

Single Input Single Output (SISO): A radio transmitter that supports one radio chain and can send and receive only a single stream of bits.

Signal to Noise Ratio (SNR): A comparison between the received signal strength and the noise floor. Typically presented in dB. For example, given a noise floor of -95 dBm and a signal strength of -70 dBm, the SNR is 25 dB.

Space-Time Block Coding (STBC): The use of multiple streams of the same data across multiple radio chains to improve the reliability of data transfer through redundancy.

Spatial Multiplexing (SM): Used with MIMO technology to send multiple spatial streams of data across the channel using multiple radio chains (radios coupled with antennas).

Spatial Multiplexing Power Save (SMPS): A power saving feature from 802.11n that allows a station to use only one radio (or spatial stream).

Spatial Streams: The partitioning of a stream of data bits into multiple streams transmitted simultaneously by multiple radio chains in an AP or client STA.

Spectrum Analysis: The inspection of raw RF energy to determine activity in an area on monitored frequencies. Useful in troubleshooting and design planning.

Spectrum Analyzer: A hardware and software solution that allows the inspection of raw RF energy.

Station (STA): Any device that can use IEEE 802.11 protocol. Includes both APs and clients.

Supplicant: In 802.1X, the device attempting to be authenticated. Also, the term used for the client software on a device that is capable of connecting to a WLAN.

Sweep Cycle: The time it takes a spectrum analyzer to sweep across the frequencies monitored. Often a factor of the number of frequencies scanned and the RBW.

System Operating Margin (SOM): The actual positive difference in the required link budget for a bridge link to operate properly and the received signal strength in the link.

Temporal Key Integrity Protocol (TKIP): The authentication and key management protocol supported by WPA systems and implemented as an interim solution between WEP and CCMP.

Transition Security Network (TSN): A network that allows WEP connections during the transition period over to more secure protocols and an eventual RSN. An RSN does not allow WEP connections.

Transmit Beamforming (TxBF): The use of multiple antennas to transmit a signal strategically with varying phases so that the communication arrives at the receiver such that the signal strength is increased.

Transmit Power Control (TPC): A process implemented in WLAN devices allowing for the output power to be adjusted according to local regulations or by an automated management system.

U-APSD: See *Automatic Power Save Delivery*.

Uncontrolled Port: In an 802.1X authentication system, the virtual port that allows only authentication frames/packets through to the network and, when authentication is successfully completed, provides the 802.1X service with the needed information to open the controlled port.

User Priority (UP): A value (from 0-7) assigned to prioritize traffic that corresponds to different access categories for WMM QoS.

Virtual Carrier Sense: The 802.11 standard currently defines the Network Allocation Vector (NAV) for use in virtual carrier sensing. The NAV is set based on the duration value in perceived frames within the channel.

Voltage Standing Wave Ratio (VSWR): The Voltage Standing Wave Ratio is the ratio between the voltage at the maximum and minimum points of a standing wave.

Watt: A unit of power. Strictly defined as the energy consumption rate of one joule per second such that 1 W is equal to 1 joule per 1 second.

Wavelength: The distance between two repeating points on a wave. Wavelength is a factor of the frequency and the constant of the speed of light.

Wired Equivalent Privacy (WEP): A legacy method of security defined in the original IEEE 802.11 standard in 1997. Used the RC4 cipher like TKIP (WPA), but implemented it poorly. WEP is deprecated and should no longer be used.

Wi-Fi Alliance: An association that certifies WLAN equipment to interoperate based on selected portions of the 802.11 standard and other standards. Certifications include those based on each PHY as well as QoS and security.

Wi-Fi Multimedia (WMM): A QoS certification created and tested by the Wi-Fi Alliance using traffic prioritizing methods defined in the IEEE 802.11e.

Wi-Fi Multimedia Power Save (WMM-PS): A power saving certification designed by the Wi-Fi Alliance and optimized for mobile devices and implementing methods designated in the IEEE 802.11e amendment.

Wireless Intrusion Prevention System (WIPS): A system used to detect and prevent unwanted intrusions in a WLAN by detecting and preventing rogue APs and other WLAN threats.

Wireless Local Area Network (WLAN): A local area network that connects devices using wireless signals based on the 802.11 protocol rather than wires and the common 802.3 protocol.

WPA-Enterprise: A security protocol designed by the Wi-Fi Alliance. Requires an 802.1X authentication server. Uses the TKIP encryption protocol with the RC4 cipher. Implements a portion of 802.11i and the older, no deprecated TKIP/RC4 solution.

WPA-Personal: A security protocol designed by the Wi-Fi Alliance. Does not require an authentication server. Uses the TKIP encryption protocol with the RC4 cipher. Also known as WPA-PSK (Pre-Shared Key).

WPA2-Enterprise: A security protocol designed by the Wi-Fi Alliance. Requires an 802.1X authentication server. Uses the CCMP key management protocol with the AES cipher. Also known as WPA2-802.1X. Implements the non-deprecated portion of 802.11i.

WPA2-Personal: A security protocol designed by the Wi-Fi Alliance. Does not require an authentication server. Uses the CCMP key management protocol with the AES cipher. Also known as WPA2-PSK (Pre-Shared Key).

Wi-Fi Protected Setup (WPS): A standard designed by the Wi-Fi Alliance to secure a network without requiring much user knowledge. Users connect either by entering a PIN associated with the device or by Push-Button which allows users to connect when a real or virtual button is pushed.

Index

2.4 GHz, 65
2.4 GHz Channels, 97
5 GHz, 65
5 GHz Channels, 100
60 GHz, 65
802.11, 62, 235
802.11 Channels, 97
802.11a, 235
802.11aa, 235
802.11aa-2012, 61
802.11ac, 235
802.11ac-2013, 61
802.11ad-2012, 61
802.11ae, 235
802.11ae-2012, 61
802.11af-2013, 61
802.11ah, 61, 236
802.11ax, 236
802.11b, 236
802.11e, 236
802.11e-2005, 77
802.11g, 236
802.11i, 236
802.11i-2004, 76
802.11n, 236
802.11r, 236
802.11r-2008, 77
802.11u-2011, 78
802.11w, 237

802.11w-2009, 77
802.1X, 63, 125, 237
802.3, 63, 237
Absorption, 16, 237
Access Point, 238
Access Point Internal vs. External Antennas, 170
Access Point Mounting Options, 171
Access Point Output Power Control, 173
Access Point PHY Support, 168
Access Point PoE Support, 170
Access Point Security Options, 178
Access Points (APs), 166
Active gain, 41
Ad-Hoc Mode, 238
AirTime Fairness, 239
Amplification, 239
Amplifiers, 54
Amplitude, 9, 239
Antenna, 239
Antenna Charts, 42
Antennas, 36
Attenuation, 13, 240
Attenuator, 240
Attenuators, 54
Authentication, 118, 121
Authentication cracking, 117

Auto-Channel Selection, 95
Autonomous AP, 240
Autonomous APs, 179
Availability, 119
azimuth chart, 43
Backward Compatibility, 155
Band Steering, 95, 240
Basic Service Area (BSA), 78, 241
Basic Service Set (BSS), 78, 241
Basic Service Set Identification (BSSID), 241
BSSID, 78
BYOD, 132
Capacity, 92, 222, 242
Captive Portal, 242
CCMP/AES, 125, 242
cell, 154
Channel, 243
Channel Selection, 97
Channel Selection Best Practices, 102
Channel Selection Mistakes, 103
Client Device Types, 188
Clients PHY Support, 196
Clients Security Features, 199
Clients Supported Channels, 197
Cloud-based APs, 180
Co-Channel Interference (CCI), 243
Coding, 26, 243
Conference Centers, 209
Confidentiality, 119
Controller-Based AP, 244

Controller-based APs, 179
Coverage, 86, 222
Data rate, 18
Data Rate Management, 96
dB, 20
dBi, 42, 245
dBm, 20, 245
decibel (dB), 20
Decibel (dB), 245
demilitarized zone (DMZ), 139
Denial of Service (DoS), 117
Deprecated Security, 130
Desktops, 192
DHCP. *See* Dynamic Host Configuration Protocol (DHCP)
DNS. *See* Domain Name System (DNS)
Domain Name System (DNS), 217
DSSS, 67, 245
dual-band, 194
Dynamic Frequency Selection (DFS), 102, 246
Dynamic Host Configuration Protocol (DHCP), 217
Dynamic Rate Switching (DRS), 153, 246
EAP, 247
EAP Methods, 129
EAP types. *See* EAP Methods
Eavesdropping, 117
Education, 210
elevation chart, 43

Emergency Response, 213
Encryption, 118, 122, 246
Encryption cracking, 117
Equivalent Isotropically Radiated Power (EIRP), 247
ERP, 70, 247
Existing Networks, 215
Extended Service Set (ESS), 78
external antennas, 53
Fast BSS Transition (FT), 158
Fast Secure Roaming (FSR), 157
Frequency, 5, 248
Frequency Bands, 65
Government, 210
Guest Access, 137
Guest Networks, 242
Healthcare, 209
Highly directional antennas, 49
Hospitality, 209
HR/DSSS, 68, 249
HT, 71, 249
IEEE, 249
IEEE Standards, 60
Independent Basic Service Set (IBSS), 249
Industrial, 211
Integrity, 119
Interference, 18
internal antennas, 53
jitter, 93
Jitter, 109, 250
Key management, 122

Key Management, 118
Laptops, 188
Latency, 93, 108, 250
Load Balancing, 95
MAC spoofing, 117
MDM, 135
Mesh BSS, 146
Mesh BSS (MBSS), 251
Mesh networks, 91
Milliwatt, 19, 251
MIMO, 149
Mobile Phones, 191
modulate, 3
Modulation, 3, 26, 251
Multi-Band, 194
MU-MIMO, 175
Neighbor Networks, 216
Network Diagrams, 220
Network Monitoring, 95
Network Services, 217
Noise, 253
Noise Floor, 18, 253
OFDM, 69, 253
Omnidirectional antennas, 44
Open System Interconnect (OSI), 157
Opportunistic Key Caching (OKC), 158
Outdoor Areas, 91, 224
Output Power, 95
passive gain, 41
Passive Gain, 254
Phase, 9

Physical Layers, 66
PoE, 255
Power over Ethernet (PoE), 218, 255
Public Wi-Fi, 214
Quality of Service (QoS), 152, 256
radiation pattern, 36
Radiation Pattern, 37
radio chain, 151
Radio Chains, 256
Radio Frequency (RF), 2, 256
Radio Resource Management (RRM), 95
Reflection, 12, 257
Remote Authentication Dial In User Service (RADIUS), 128
Remote Authentication Dial-In User Service (RADIUS), 256
Request to Send/Clear to Send (RTS/CTS), 259
Retail, 211
RF. *See* Radio Frequency (RF)
RF Behaviors, 11
RF cables, 54
RF Characteristics, 5
RF connectors, 54
RF Radiation, 37
RF Reception, 40
Roaming, 157
RSSI, 21, 257
S1G, 75
Security Constraints, 226
Semi-Directional Antennas, 47

Service Set Identifier (SSID), 78, 260
Signal Strength, 18, 109
Signal to Noise Ratio (SNR), 260
signal-to-noise ratio (SNR), 18
Single-Band, 194
SISO, 149, 260
Small Office/Home Office (SOHO), 214
SNR. *See* Signal-to-Noise Ratio (SNR)
Social engineering, 118
SOHO, 214
spatial streams, 26, 152
Specialty Devices, 193
Standard Enterprise Offices, 208
Standard Office Space, 87
Standards Development, 60
Sub-1 GHz (S1G), 65
SU-MIMO, 175
Tablets, 191
Temporary Deployments, 213
Throughput, 18
Throughput requirements, 93
TKIP/RC4, 123
TVHT, 74
Utilization, 93
VHT, 73
Virtual Local Area Network (VLAN), 138
Voice over WLAN (VoWLAN), 103
Warehouse, 89
Watt, 19, 262

wave, 4
Wavelength, 6, 262
Wi-Fi Alliance, 263
Wi-Fi Implementations, 215
Wireless Local Area Network (WLAN), 263
WLAN Scanners, 22
WLAN Security, 116
WNMS-managed APs, 181
WPA, 123
WPA2, 123
WPA2-Enterprise, 128, 263
WPA2-Personal, 128, 264

Made in the USA
Monee, IL
03 August 2021